PREDICTING

The Adventure Begins

Robert Ellis

PREDICTING STORMS

The Adventure Begins

Third Edition

Robert Ellis

Published in Australia in 2018 by Goldener-Parnell Publishing
Reprinted 2020
Third edition 2024

Email: rob@worldstormcentral.co
Website: http://www.worldstormcentral.co

ISBN 9780648107200 (paperback) – 1st ed.
ISBN 9780648107224 (paperback) – 2nd ed.
ISBN 9780648107286 (paperback) – 3rd ed.
ISBN 9780648107255 (hardback) – 2nd ed.
ISBN 978-0-6481072-7-9 (hardback) – 3rd ed.
ISBN 9780648107248 (Ebook - EPUB)
ISBN 9780648107262 (Ebook - Adobe PDF)

NATIONAL
LIBRARY
OF AUSTRALIA

A catalogue record for this book is available from the National Library of Australia

Disclaimer

The author has made every effort to ensure the accuracy of the information within this book was correct at the time of publication. The author does not assume and hereby disclaims any liability to any party for any loss, damage or disruption caused by errors or omissions, whether such errors or omissions result from accident, negligence, or any other cause.

This book is dedicated to the memory of my parents.

Contents

ACKNOWLEDGEMENTS ix

TABLE OF FIGURES xi

TABLE OF TABLES xiv

INTRODUCTION 1

Ground-breaking Storm Prediction 1

A Foundation Work 2

Nature's Law of Storms 2

Predicting Storms is Easy and It's Fun 2

Barometers 3

More Reasons to Use a Barometer 4

Barometer Rules to Get Started 5

All Types of Storms Covered 6

Sea and Surf 6

Why Expect Stronger Storms in a Warmer World? 6

The Slow Train is Coming Around the Bend 7

Advances in Storm Early Warning 7

Visit our Website 7

1. RULES FOR STORMS 9

How does a Storm Develop? 9

Rule for Thunderstorms 11

The Lightning Threshold 21

Weather Bomb 22

Gale	23
Strong Wind	25
Severe Thunderstorm	25
Supercell Thunderstorms	32
Tornadoes	41
Hailstorms	49
Cyclones/Typhoons/Hurricanes	51
Firestorms	54
2. SEA AND SURF	63
Early Warning of Storms	63
Secret of the Swell	66
Estimating Hours Before a Hurricane Makes Landfall	70
Estimating the Wind	70
Where you are in Relation to Low or High-Pressure Systems	71
Crossed Winds Rule	71
Gusts	72
3. PREDICTING RAIN	73
An Altimeter Can Indicate Stormy Weather	74
Rapid Pressure Fall Foretells Storm or Rain	74
4. CLOUD SEQUENCES	75
Sequence of Cloud Arrival for Storms	75
Cloud Sequence for Cyclone/Hurricane/Typhoon	81
Determining Cloud Level	82
Calculate the Cloud Base Height	82
Cloud Sequences in a Nutshell	83
5. WHY A SMALL INCREASE IN CO_2 CRITICALLY AFFECTS CLIMATE	85
No Slow Down in Global Warming	87
Why is a Change in the Earth's Global Average Temperature a Big Deal?	88

Contents

Evidence for Rapid Climate Change is Compelling   89

Stronger Storms in a Warmer World   90

World Population Growth is the 'Elephant in the Room'   92

How was Global Warming Discovered?   92

6. WINDSTORMS   95

7. SNOWSTORMS   97

CONCLUSION   99

APPENDIX 1:   WEATHER LORE SAYINGS   105

APPENDIX 2:   DERIVATION OF THE LAW OF STORMS   107

APPENDIX 3:   TYPES OF BAROMETERS   109

APPENDIX 4:   STORM RECORDS   113

APPENDIX 5:   DOUBLING THE CO_2 CONTENT   115

APPENDIX 6:   THE THUNDERSTORM RULE IN A NUTSHELL   116

ISLAND IN THE STORM: A SHORT STORY   117

REFERENCES   125

PERMISSIONS   131

ABOUT THE AUTHOR   137

Acknowledgements

Acknowledgement is given to my parents for their dedicated support in all that I have done. Harry Ellis, my father, taught me that 'if a job is worth doing it is worth doing well.' You learn what you live, and he shared with me a passion for observing Australia's natural environment with its droughts and flooding rains. He had his own weather station. My brother, Garry Ellis, had as keen interest in the weather. My mother, Jean Ellis, was a constant support throughout my life and with all those years of research into storms. Thanks to Luke Ellis, who reminded me of the good swell for surfers that arrives from a distant storm.

Special thanks are given to U.S. Department of Commerce National Oceanic and Atmospheric Administration, and to the Bureau of Meteorology Australia, especially to their librarians and photographers, for their generosity in making available their resources. Thanks to the librarians at the Sutherland Shire Library for their assistance over the years with my research.

Thanks to Dr Jeff Masters, Director of Meteorology at wunderground.com and Dr Greg Holland for their assistance. Thanks to Jane Rothman and Dr Max Humphreys for their support.

Acknowledgement is given to all those who have worked on the first edition of the book, especially Gail Tagarro of editors4you, managing editor, who ensured, amongst so many other things, that the book was thoroughly referenced and the content was easily read by everyone. Kirsty Ogden from Brisbane Self Publishing Service is acknowledged for her excellent typesetting and cover design, as well as her assistance with the second edition of the book.

Thanks to Josh Waghorn, Bronwyn Melville and Reuel Mantos for graphic design work and Nurali Prasla's team for programming the Thunder & Bushfire Storms app.

Table of Figures

Figure 1:   Thunderstorm heat engine cycle.   10

Figure 2:   Lightning bolt during storm, Tallahassee, Florida,   12
            U.S.

Figure 3:   Thunderstorm over Watson Lake, Prescott, Arizona,   14
            U.S.

Figure 4:   Stylised barograph of thunderstorm signature.   15

Figure 5:   Intense cloud to ground lightning over southern   16
            Lake Michigan, Chesterton, Indiana, U.S.

Figure 6:   Updraft tilts with wind shear.   17

Figure 7:   Twin towering cumulus.   19

Figure 8:   Cumulonimbus with magnificent anvil.   20

Figure 9:   Lightning before rain. From back yard in Rochester,   21
            New York, town of Greece.

Figure 10: Gale-force winds lash huge ocean waves into a   24
            violent stretch of water. Drake Passage between
            Atlantic Ocean and Pacific Ocean.

Figure 11: Pressure graph (mb) for a Severe Thunderstorm.   26

Figure 12: Cold front with squall line. Northern Adriatic Sea,   28
            Italy.

Figure 13: Sailboat approaching squall line, Pamlico Sound,   29
            North Carolina, U.S.

Figure 14: Overshooting top cumulonimbus cloud.   30

Figure 15: Rain shaft left, rain-free base of severe thunder-   31
storm right. Key West, Florida, U.S.

Figure 16: Severe Thunderstorm at Era Beach south of Sydney.   32
The wall cloud (pedestal cloud) is lowered beneath
the base of a cumulonimbus cloud. Picture also
shows rain-free base that increases storm's longevity.

Figure 17: Aerial view of a supercell thunderstorm. Photo   34
taken looking northeast over eastern Kansas, U.S.

Figure 18: Intense updrafts produce a rain-free cloud base in a   35
supercell.

Figure 19: Stylised supercell thunderstorm.   36

Figure 20: Pressure profile – supercell with deadly tornado.   37

Figure 21: Pressure profile – supercell with tornado.   38

Figure 22: Stylised barographs of tornadic supercell thunder-   40
storm.

Figure 23: Waterspout in the Gulf of Mexico photographed   42
from the NOAA ship Rude. South of Cameron,
Louisiana, Gulf of Mexico.

Figure 24: Twin violent (EF4) tornadoes, Wisner, Nebraska,   44
U.S.

Figure 25: Signature shape shows tornado proximity.   46

Figure 26: Concave down curve.   47

Figure 27: Barograph of a tornado (Jensen 2010).   48

Figure 28: Lightning shoots up updraft & anvil of tornadic   50
supercell at night, with car light trails.

Figure 29: Thunderstorm heat engine cycle.   51

Figure 30: Stylised barogram showing 6-hour steady pressure   53
interval.

Figure 31: Hurricane viewed from satellite.   54

Figure 32: How firestorms form.   57

Figure 33: Stylised barographs of typical storm signatures.   58

Figure 34: Thunderstorm quick basics.   59

Figure 35: Predicting storms.   60

Figure 36: Barometer Rules   61

Figure 37: Swell lines in the Pacific.   65

Figure 38: Swell from a distant storm apparent on the sea surface.   67

Figure 39: Swell.   69

Figure 40: Virga viewed SW from Flat Top Mountain, North Carolina, U.S.   72

Figure 41: Cirrus at sea.   76

Figure 42: Cirrocumulus cloud, Michigan, Grand Rapids, U.S.   77

Figure 43: Sun halo and cirrostratus clouds at sunset.   79

Figure 44: Mackerel sky of altocumulus clouds over eliptic crater, Erta Ale volcano.   80

Figure 45: Data showing no recent slowdown in global warming.   87

Figure 46: Dust storm approaching Stratford, Texas, U.S.   96

Figure 47: The 2016 snowstorm in Washington D.C. ranked as a category 4 storm on the NESIS scale.   98

Figure 48: Mercury barometer.   110

Figure 49: Storm cloud near Bundeena, Australia.   117

Figure 50: Lifeguard chair on beach.   122

Table of Tables

Table 1:   Thunderstorms in a nutshell.   22

Table 2:   Results obtained by applying Law of Storms   23
            formula for a gale.

Table 3:   Typical Category 2 hurricane/Category 3 cyclone   68
            approaching coastal area.

Table 4:   Hours before waves make landfall.   69

Table 5:   Estimating the wind—at a glance.   70

Table 6:   Barometer rules for rain.   73

Table 7:   Cloud sequences.   75

Table 8:   Cloud sequences in a nutshell.   83

Introduction

Ground-Breaking Storm Prediction

Currently, 13 minutes lead time is all that can be definitely given before a tornado strikes. The Tornado Early Warning Rule presented here is ground-breaking and will save lives. You can now have *at least 5 hours early warning of a deadly tornado from its rigid straight-line signature on a barograph (see page 37). Being able to have at least 5 hours early warning of a deadly tornado is a breakthrough.* Good weather stations usually have a barograph showing a trace of the barometric pressure over time. You will be able to detect any storm using a simple barometer long before it is visible to radar or satellite. (See Tornado Early Warning Rule, page 36.)

Doppler radar can detect rotation within a tornadic supercell thunderstorm but can only give up to 20 minutes early warning of a tornado. The existence of a so-called 'Tornado Vortex Signature' on radar does not even guarantee that a tornado will actually occur. While satellite information can be useful, a satellite views a cloud from above and cannot see, for example, a tornado below. The time delay in receiving a satellite picture can be up to 30 minutes, which does not allow adequate early warning of severe weather.

Being able to use your barometer can be a real game-changer in predicting storms. Up to as many as 500,000 people worldwide can die in a large storm in a single year. *Your barometer will give you at least*

1

24 hours early warning of an approaching hurricane making landfall.
(See Hurricane Early Warning Rule, page 53). You can have early
warning of a bush firestorm (see page 55).

A Foundation Work

The first edition of this book stands as a foundation work on storm
prediction and barometer rules.

'In a sense this (book) is probably the first comprehensive compila-
tion of rules for ordinary storms.' (McGuire, 2019, p17).

This second edition consolidates that foundation. The barometer
rules here rely entirely on the laws of physics on which we all depend. They were compared with 'centuries of accumulated lore that
sailors used to live by to see what could be learned as well.'
(McGuire, 2019, p16).

Nature's Law of Storms

In 2014, Robert Ellis revealed Nature's law governing storms and gales.
He transformed a well-known physics equation by calculus into
a simple law of storms. This Law of Storms is expressed as a
mathematical formula based on a law of physics, not on chance.
In our lives, we rely entirely on the laws of physics.

Predicting Storms is Easy and it's Fun

Atmospheric (or barometric) pressure is caused by the weight of air
pushing down on us. Weather changes the atmospheric pressure.
Pressure is measured in millibars (mb). An ordinary thunderstorm
requires the last 3 hours pressure fall of 4 mb or more to below 2009
mb. This is called The Thuderstorm Rule that we will explain later.
If this storm threshold is passed without the storm onset, storm development will deepen until eventual onset. The Thunderstorm Rule is
a necessary condition for all storms, but a Severe Thunderstorm,
Supercell Thunderstorm, Cyclone/Hurricane/Typhoon, Deadly Tor-

nado, Bush Firestorm and Fire Tornado require the additional conditions given here to uniquely identify them.

> **When the barometric pressure falls 4 mb or more in the last 3 hours to below 1009 mb, a storm is approaching.**

This book shows you how to accurately predict the storm type and size. You will be able to confirm your prediction by observing the sequence of clouds arriving from the distant storm.

You don't need a traditional forecaster, a satellite or supercomputer. That would be like using a huge sledgehammer to drive in a little nail.

Barometers

A barometer measures the subtle air pressure changes caused by weather. Some popular smart phones have a built-in barometer. Or you can buy a good affordable barometer online. The smart phone with a barometer sensor and a barograph is a most precious standalone device. It is not affected by poor reception or power outages. You can use it at sea and in remote areas where there is no radar coverage. It is more economical if it uses pressure sensors and does not require the internet. Barometer rules are essential to interpret our observations. *The barometer gives real-time pressure data at the immediate location, and tells of approaching weather, often long before it is visible, providing an early warning.* For example, a barometer can provide at least 5 hours early warning of a *tornadic* supercell thunderstorm, long before it is visible to radar or satellite. Tornadic simply means the supercell thunderstorm will produce a tornado. This book gives the early warning barometer rule.

An approaching storm can be predicted by tracking barometric pressure and its rate of change. General weather is affected by many factors, including pressure, temperature, wind speed, and humidity. Fortunately, a storm's maximum wind speed (intensity) depends *entirely* on its central pressure. *That is why the barometer is such a powerful instrument for predicting storms.*

More Reasons to use a Barometer

When the traditional weather forecaster says there is a 60% chance of a storm, it still may not happen, or it may occur elsewhere. Fortunately, you can accurately predict a storm in your immediate vicinity using the barometer rules given here. Whereas traditional forecasts are often generalised to cover a much wider area, giving no mention of individual storms, a barometer can determine the occurrence of a storm in your immediate location or close to it. Severe Thunderstorms, and some storms on land and at sea, are often localised and it can be hard for weather services to predict their exact location; even technology cannot detect every storm. It also takes time for weather services to obtain storm-spotter reports to verify information. Weather supercomputer models are becoming less reliable and the weather forecasts less predictable as the Earth's average global temperature increases (Global Warming). Historical weather patterns on which forecasts traditional rely are becoming increasingly unreliable. Fortunately, barometer rules given here remain reliable. ***Look for an online weather station (with a barograph) near your location.***

If you want to know whether you can walk to work, or if there will be a storm in your vicinity within an hour or two, a barometer is invaluable.

According to the Bureau of Meteorology (BOM) in Australia 'The detailed warnings are usually issued without much lead-time before the event.' They have difficulty in forecasting 'the precise location and movement of severe storms before they have started to develop' Extracted 20 February 2024. ***A barograph app with a pressure sensor (or a traditional barometer) and the simple rules in this book allow ordinary people to discern the type of storm with much lead-time before it onsets.***

Barometer Rules to Get You Started

You may wonder, when will you be safe from storms?

Safe conditions are:

- If the pressure remains above 1009 mb, there can be no storms.
- If the pressure remains above 1005 mb, there can be no severe thunderstorms, including Supercell Thunderstorms or Tornadoes.
- If the pressure remains above 1000 mb, there can be no cyclones, hurricanes or typhoons.
- If the pressure falls less than 3 mb in the last 3 hours, there can be no storms.

When the pressure starts falling at more than 1.0 mb / hour a storm may be approaching.

Apply **The Thunderstorm Rule** to be certain:
(i.e. when barometric pressure falls 4 mb or more in the last 3 hours to below 1009 mb, an ordinary thunderstorm is approaching). If there is no storm onset it means a *Severe Thunderstorm* is developing.

◄──────────────────────────────────────►

You can have at least 5 hours early warning of a deadly tornado from its rigid straight-line signature on a barograph.

You can have 24 hours early warning of a Hurricane in the tropics when the pressure has been almost steady for 6 hours.

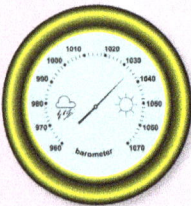

Try the above barometer rules now to get started.
Use your own barometer or access a graph of pressure against time (barograph) in an app with pressure sensor. Marine Barograph app can be used on land or at sea. Barometer readings need to be taken regularly, usually every 6 hours but even more frequently at sea or especially when the pressure starts falling at more than 1.0 mb / hour.
When the barometer is steady the current weather you are observing will persist.
www.worldstormcentral.co

All Types of Storms Covered

All types of storms are covered in this book, including thunderstorms, Severe Thunderstorms, tornadic supercell thunderstorms, cyclones, hurricanes, typhoons, extratropical cyclones, tropical storms, tornadoes, bush firestorms, fire tornadoes, weather bombs, windstorms, dust storms and snowstorms.

Sea and Surf

Sea and surf are another important part of this book. The barometer rules apply to storms on land *or* at sea. Barometer readings need to be taken regularly, usually every 6 hours but even more frequently at sea or especially when the pressure starts falling at more than 1.0 mb/h. *When the barometer is steady the current weather you are observing will persist.* There will be other times that allow less frequent monitoring of barometric pressure. A wind speed greater than 15 knots (28 km/h) can capsize a small boat. We will show that a minimum 6 mb fall in barometric pressure over 3 hours will produce a strong wind of 29.3 knots (54.2 km/h). There is a *Marine Barograph* app in the Apple App Store and Google Play Store that can be used on land or at sea. If you have a second or spare iPhone 6 or newer device you can use them as a standalone barograph for continual monitoring.

Challenge: The planet Venus has the highest surface temperature in the Solar System yet it is not the closest planet to the sun. Why is it so? Discover the reason for global warming. See answer given in Chapter 5.

Why Expect Stronger Storms in a Warmer World?

A small increase in carbon dioxide content in the atmosphere makes a critical difference to the Earth's climate. This is explained in more detail on page 90.

The Slow Train is Coming Around the Bend

"Changes in ocean systems generally occur over much longer time periods than in the atmosphere, where storms can form and dissipate in a single day. Interactions between the oceans and atmosphere occur slowly over many months to years, and so does the movement of water within the oceans, including the mixing of deep and shallow waters." (https://www.epa.gov/climate-indicators/oceans Extracted 26 August 2023)

Even if greenhouse gas emissions were stabilized tomorrow it would take at least 30 years for the oceans to adjust to changes in the atmosphere and the climate that have already occurred. The oceans today reflect the climate and global warming of at least 30 years ago. *How severe will be the storms of our grandchildren!*

Advances in Storm Early Warning

Severe Thunderstorm Early Warning Rule, page 27

Supercell Thunderstorm Early Warning, page 35

Tornado Early Warning Rule, page 36

Visit our Website

The website associated with this book is www.worldstormcentral.co

1

Rules for Storms

How does a Storm Develop?

Warm moist air is the *fuel* of the storm heat engine. When it rises, it cools and condenses, releasing heat. Storm development starts when the sun heats up the ground or ocean surface, causing evaporation. A column of warm, moist air stores latent heat from this evaporation and rises, reducing the air pressure at the base of the column—this is called the central pressure core. As the warm, moist air rises, it cools, condenses into clouds and releases latent heat fuel, which accelerates the updraft (see below) and further reduces the central pressure. The heat engine chugs away, accelerating latent heat release, with a corresponding further central pressure fall. Moist air then rushes into the central lower pressure core.

This process loop continues, with the updraft speed at the freezing level eventually passing the threshold for a storm.

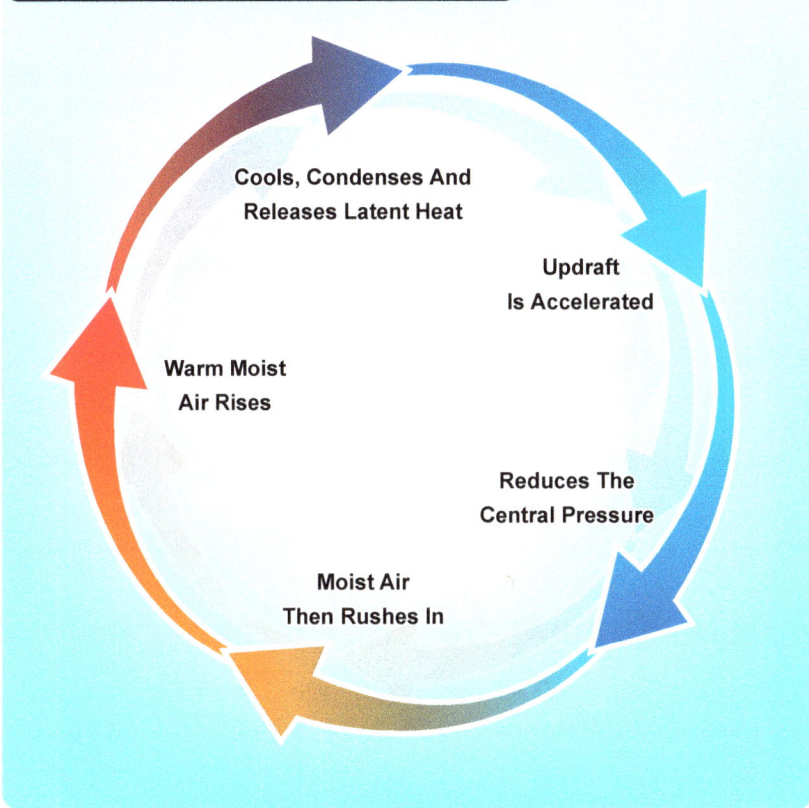

Figure 1: *Thunderstorm heat engine cycle.*

Updraft: The updraft is observed as dark cloud where warm unstable air rises. The inflowing updraft will blow into your back if you stand facing a storm.

Downdraft: The downdraft, usually white or clear, is where rain-cooled air sinks. The rear flank downdraft can be identified in a supercell thunderstorm as a 'clear slot' in the storm cloud. Outflow winds blow away from a storm. They tend to be cold with condensation. Turbulent winds of gusts fronts or shelf clouds move away from a storm.

"**Supercells are storms characterised by** *updrafts* **that can reach over 100 mph (160 km/h) and produce giant hail with strong or even violent tornadoes. Their downdrafts can produce downbursts/outflow winds greater than 100 mph (160 km/h), and can pose a high threat to life and property (NOAA n.d.).**"

Rule for Thunderstorms

By tracking barometric pressure and its rate of change, and observing that the barometric pressure is below its ceiling value and the rate of change exceeds a threshold, an approaching storm can be predicted. A ceiling value means that for a storm to occur, the pressure must fall below that value. As above, for an ordinary thunder-storm, the ceiling value will be 1009 mb and the rate of pressure fall at least 4 mb in the last 3 hours.

The ceiling value for a Severe Thunderstorm is 1005 mb.
The pressure fall to below 1005 mb for a Severe Thunderstorm can take 8 hours to occur.

Figure 2 (on following page): *Lightning bolt during storm, Tallahassee, Florida, U.S.*

We will now apply the Law of Storms.

Air moves as wind when there is a pressure difference between two locations. Wind flows towards the lower pressure centre of a storm. The Law of Storms tells us that a central pressure fall (1009 to 1006 mb) in the last 3 hours in the formation of a storm will increase wind speed from 9.8 to 25.1 km/h:

Pressure (mb)	Wind speed (km/h)
1009	9.8
1006	25.1

A wind speed of 25.1 km/h (15 mph) is the threshold for a thunderstorm.

Therefore, a minimum of 3 hPa fall in barometric pressure over 3 hours will produce a thunderstorm. This smallest recorded thunderstorm, with a 3 hPa fall in barometric pressure over 3 hours (1009 hPa to 1006 hPa), was recorded at Middlebury, Vermont, US on 7 June 2011.

A 4 hPa fall in barometric pressure over 3 hours can be used to provide a margin of comfort for prediction purposes. It can be shown that a minimum of 4 hPa fall in barometric pressure over 3 hours will increase the wind speed to 29.2 km/h. **A storm also typically requires the barometric pressure to be less than 1009 hPa (or mb).** If barometric pressure has fallen 4 mb, or more, to below 1009 mb in the last 3 hours, a storm is approaching.

Figure 3 (on following page): *Thunderstorm over Watson Lake, Prescott, Arizona, U.S.*

Storm signature on barographs

You can detect any storm using a simple barometer, long before it is visible to radar or satellite. A barograph is a graph of pressure over time. The characteristic shape of the barograph for each type of storm (its storm signature) is a distinct way of uniquely identifying the type and size of a storm. Recognising the storm signature can provide early warning of a thunderstorm or Severe Thunderstorm (up to 2 hours), Tornadic Supercell Thunderstorm (at least 5 hours) and Hurricane (24 hours).

If you look at a barograph, a coming storm will show a characteristic pressure dip. It often looks like a tick or 'v' shape with a short steep rise or jump in pressure immediately after the pressure dip. This is the storm signature. The steep pressure jump can on occasion exceed 1 mb per minute. The pressure jump is caused by the local high pressure that usually accompanies the storm. On completion of the pressure jump, the temperature may have fallen by about 8.35°C (15°F) and precipitation may have commenced. You will still need the barometer rules to confirm that it is a storm. That is, whether the barometric pressure has fallen 4 mb, or more, to below 1009 mb in the last 3 hours.

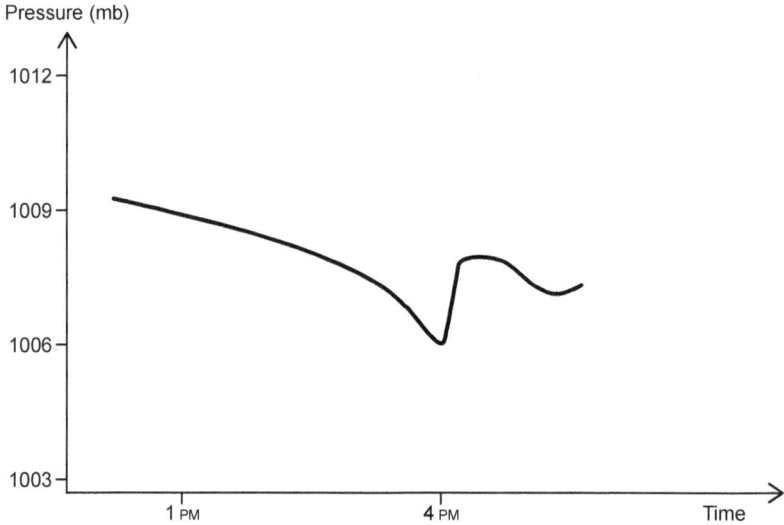

Figure 4: *Stylised barograph of thunderstorm signature.*

Other storm events on barographs

A barograph alerts you to every significant event during a storm. These events are reflected in the shape or features of the pressure curve. When the pressure varies significantly from its established trend, it signals an important change in storm behaviour. You can see when a storm will give a jump in the lightning flash rate, produce hail or spawn a tornado.

Criteria for a thunderstorm

An ordinary thunderstorm requires:

- barometric pressure of less than 1009 mb, and
- last 3 hours pressure fall of 4 mb or more.

This is called the *Thunderstorm Rule*.

> *The Thunderstorm Rule can be used to detect any storm, but further rules need to be applied to uniquely identify the type and size of a storm.*

15

Typically, ordinary thunderstorms can last about an hour. Storm coverage can extend to a diameter of 20 km. A thunderstorm is initially produced by rising, warm, moist air (updraft), while cold, sinking air (downdraft) increases as the storm matures. The updraft drives the storm. As the storm matures, the colder downdraft gradually extinguishes the warm updraft and the storm itself. Updrafts that are almost vertical and not displaced from the downdraft cause the rain to fall into the updraft shaft and reduce the storm's longevity.

Figure 5 (below): *Intense cloud to ground lightning over southern Lake Michigan, Chesterton, Indiana, U.S.*

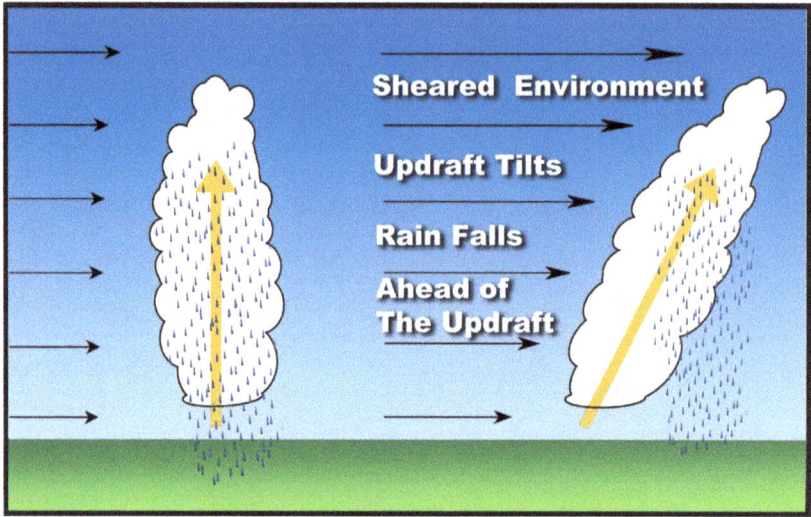

Figure 6: *Updraft tilts with wind shear.*

Vertical wind shear occurs when the horizonal wind blows faster as one ascends, so the updraft shaft becomes tilted. The *higher* the wind shear, the more the updraft shaft is tilted, causing the rain to fall further away from the shaft and avoiding early extinction of the storm.

Ordinary thunderstorms are frequent in the tropics, where there are usually only moderate variations in temperature and there is low wind shear. Updrafts are almost vertical and are not displaced from downdraft, causing the rain to fall into the updraft shaft, reducing the storm's longevity. *The updraft is far more tilted with height in Severe Thunderstorms, greatly increasing the storm's longevity.*

> Every thunderstorm cloud has a core region, a spreading anvil top, and an inflow-outflow region. The core is that part of the cloud where sustained strong updraughts of relatively warm and moist air condense to produce rain, hail and/or snow (collectively known as precipitation) and associated downdraughts. Underneath the core we see a rain curtain, whilst above it the tallest part of the thunderstorm can be found. The dark flat cloud base that extends away from the core (usually to the west or north) is called the flanking line or rain-free base, along which air

17

fuelling updraughts into the thunderstorm rises in successive cumulus towers. (Reproduced by permission of Bureau of Meteorology, © 2017 Commonwealth of Australia.)

Thunderstorm life cycle

Ordinary thunderstorms can last from 30 to 75 minutes. Knowing what stage a thunderstorm has reached enables you to predict further.

Thunderstorms have three distinct stages:

Developing stage (cumulus or growth)

- Updraft dominates as warm moist air condenses into growing, towering cumulus clouds.

- There may be a little precipitation in the upper portion of the cloud and occasional lightning.

- This stage usually lasts about 15 minutes. The next stage—mature—can last up to twice as long.

Figure 7 (on following page): *Twin towering cumulus.*

Mature stage

- There is both an updraft and a downdraft. Heavy precipitation and a cool downdraft dominate. The leading edge of the downdraft reaches the ground in a gust front.

- Downdrafts can cause the temperature to fall rapidly, on average around 8.35°C (15°F) due to evaporative cooling.

- There is frequent lightning and strong winds.

- An anvil (anvil-shaped cloud—see photograph below) will form at the top of the outflow when the top spreads out against the bottom of the stratosphere.

Dissipation stage (winding down)

- Downdraft dominates this stage, which lasts 30 minutes.

- The gust front shifts much further away from the thunderstorm and almost completely cuts off the inflow of warm moist air.

Figure 8 (below): *Cumulonimbus with magnificent anvil.*

The Lightning Threshold

Lightning occurs as follows:

- Thunderstorm inflow rises through the cumulus cloud as an updraft

- The updraft carries water droplets above the deep convective freezing level, some forming ice crystals or larger super-cooled ice crystals joined together. Droplets and crystals collide and fall at different speeds, producing lightning.

The threshold cumulus updraft speed of 7 m/s (25 km/h) has been suggested by Del Genio (2007) for the occurrence of lightning.

Lightning occurs when the increase in the rate of pressure fall causes the updraft speed to exceed the threshold for lightning. This lightning threshold is exceeded by the wind speed (29.2 km/h) of a thunderstorm formed when the barometric pressure has fallen 4 mb or more to below 1009 mb in the last 3 hours.

Figure 9 (below): *Lightning before rain. From back yard in Rochester, New York, town of Greece.*

Thunder arrives 3 seconds per km (5 secs per mile) after lightning.

Thunderstorms in a Nutshell

- Warm moist air rises (updraft) causing pressure to fall and replacement air to rush in.

- Moist air cools, condenses and releases latent heat, accelerating the updraft.

- Above freezing level, droplets and ice crystals collide and fall. Lightning occurs when updraft speed exceeds 7 m/s.

- A cool downdraft carrying rain develops as the storm matures.

Table 1: *Thunderstorms in a nutshell.*

Weather Bomb

We will now apply the Law of Storms to weather bombs. A weather bomb is a low-pressure area of an extratropical cyclone (i.e. a cyclone outside of the tropics) where the pressure is rapidly falling.

Weather bombs occur when the central pressure of a depression at 60° latitude decreases by 24 hPa or more in 24 hours (Sanders 1980). That is, the barometric pressure falls steadily by about 1mb (or 1 hPa) per hour within a 24-hour period. The winds become stronger as the barometric pressure falls. The process is known as 'explosive cyclogenesis'. It can be caused by the rapid deepening of a low-pressure area associated with an extratropical cyclone. An extratropical cyclone is characterised by warm, moist air from the low-pressure area flowing up and over cold air.

Consider this example: barometric pressure falls from 1009 hPa at a rate of 1 hPa per hour for 24 hours. It will fall to 986 hPa. Applying the Law of Storms, the resultant wind speed will be 77.1 km/h. This example shows that even the smallest weather bomb will produce a wind of gale-force strength.

Gale

To obtain the rule for a gale by applying the Law of Storms, you need to observe a minimum 10 hPa rise *or* fall in barometric pressure over 3 hours.

See the example below:

Pressure fall (mb)	Wind speed increase (km/h)
1002	37.4
992	64.2

Table 2: *Results obtained by applying Law of Storms formula for a gale.*

Also, a rise or fall in barometric pressure of more than 3 mb per hour will produce a gale at sea. Rule gives an **early warning** of a gale at sea. Land winds are about 2/3 of the offshore wind strength.

Figure 10 (on following page): *Gale-force winds lash huge ocean waves into a violent stretch of water. Drake Passage between Atlantic Ocean and Pacific Ocean.*

Strong Wind

Similarly, it can be shown that a minimum of 6 hPa fall in barometric pressure over 3 hours will increase the wind speed to 54.2 km/h or 29.3 knots from dead calm. Therefore, *a minimum of 6 hPa fall in barometric pressure over 3 hours will produce a strong wind*. The rule for a strong wind also applies to either a rise or fall in pressure.

Severe Thunderstorm

The Thunderstorm Rule enables you to predict a storm, which may be an ordinary thunderstorm or a Severe Thunderstorm, providing up to 2 hours *early warning*. Or it could predict a cyclone/hurricane/typhoon, providing about 5 hours *early warning*. This is a condition necessary for *all* storms, but a Severe Thunderstorm and a cyclone/hurricane/typhoon require additional conditions to uniquely identify them.

Storm coverage for an ordinary thunderstorm can extend to a diameter of 20 km. The coverage of a Severe Thunderstorm, such as a tornadic supercell thunderstorm, can extend to a diameter of 40 km, twice that of an ordinary thunderstorm.

An ordinary thunderstorm requires both of the following:

- barometric pressure of less than 1009 mb
- last 3 hours pressure fall of 4 mb or more.

A Severe Thunderstorm is not simply a thunderstorm that is 'severe'. It is a category in itself, with special criteria.

Even though the pressure fall can be spread over 12 hours, a Severe Thunderstorm will still have the last 3-hour interval during which there will be a minimum 4 mb fall in pressure. Sometimes, the pressure fall of 8 mb or more can occur in 8 hours or less, giving an even stronger Severe Thunderstorm.

A Severe Thunderstorm requires all the following:

- barometric pressure of 1005 mb or less

- last 3 hours pressure fall of 4 mb or more

- last 12 hours pressure fall of 8 mb or more.

A Severe Thunderstorm produces one or more of the following:

- wind gusts of 90 km/h (56 mph) or greater

- hail of 2 cm (¾ inch) diameter or greater

- flash floods

- tornadoes.

Only a few actual or real storms ever reach their maximum sustainable wind speed. This is one of the reasons a 20% adjustment in wind speed for ordinary thunderstorms is applied (refer to Equations 1, 2 and 3 on pages 107 and 108). However, this adjustment can be greatly reduced when considering Severe Thunderstorms close to reaching their full potential. Applying the Law of Storms gives a sustainable wind speed for a severe storm of 56 km/h (35 mph). Severe Thunderstorms often generate strong wind gusts when rain and hail drag down the surrounding air. Evaporation further cools it, accelerating the downdraft. Gusts of at least 90 km/h (56 mph) are produced.

The pressure graph below was recorded on 9 April 2017 at Gymea Bay, Australia.

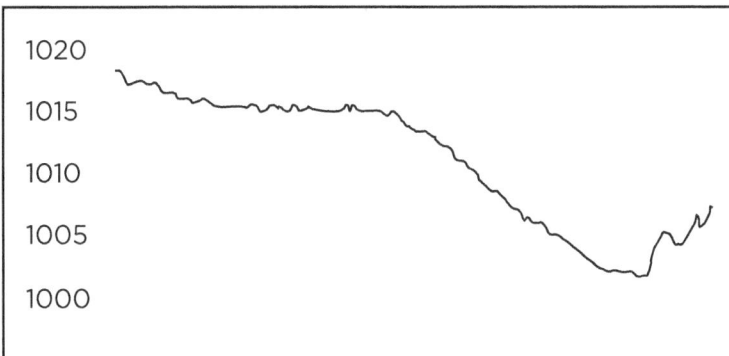

Figure 11: *Pressure graph (mb) for a Severe Thunderstorm.*

The barograph of a Severe Thunderstorm will show the pressure falls steadily for at least 8 hours to below 1005 mb. When a Severe Thunderstorm is ready to end in the usual way, the pressure continues falling to a minimum and then jumps, the storm then finally onsets within 2 hours. The jump can be almost vertical on the barograph trace for a **_hailstorm._** When the threshold for an ordinary thunderstorm (see _The Thunderstorm Rule,_ page 5) is passed without storm onset a Severe Thunderstorm is developing.

Severe Thunderstorm Early Warning Rule

> _You have at least 5 hours (and possibly up to 9 hours) Early Warning of a Severe Thunderstorm when_
>
> (a) _the pressure falls 4 mb or more in the last 3 hours to below 1005* mb without any sign of storm onset; and_
> (b) _there is an ongoing steady pressure trend._

*It is the rate of pressure fall (i.e. **_4 mb or more in the last 3 hours)_** **_without any sign of storm onset_**, that is critical for early warning purposes. It is easy to see whether after 8 hours the pressure will eventually fall below 1005 mb when after 3 hours, without any signs of storm onset, the barograph displays an ongoing steady falling pressure trend.

Squalls

Squalls can be especially dangerous at sea, damaging sails or capsizing sea vessels.

A squall is:

> a strong wind characterized by a sudden onset in which the wind speed increases at least 16 knots and is sustained at 22 knots or more for at least 1 minute, or (in nautical use) a severe local storm considered as a whole, that is, winds and cloud mass and (if any) precipitation, thunder and lightning (US Department of Commerce National Oceanic and Atmospheric Administration 2009).

Figure 12 (on following page): _Cold front with squall line. Northern Adriatic Sea, Italy._

Squall lines

Severe Thunderstorms can form into a squall line. Squall lines consist of many individual storm cells and consequently can last for hours, with sudden strong wind gusts, concentrated heavy rain, snow or sleet. Each storm cell usually lasts about 40 minutes and one storm cell at a time will tend to dominate. A squall line which can be hundreds of kilometres long will commonly form along or ahead of a cold front (McIlveen p.357). The cold air wedges the warmer air upwards.

High wind shear where the wind speed increases with height maintains the separation of the updraft and downdraft, increasing the storm's longevity. *A squall line in the tropics can be a harbinger of an approaching hurricane.*

Figure 13 (below): *Sailboat approaching squall line, Pamlico Sound, North Carolina, U.S.*

The squall line can move ahead of the hurricane, bringing hail and gusts. Away from the tropics, barometric pressure falls steadily until the cold front arrives, and rises sharply and then steadily once it passes. The pressure will reverse direction within an hour when a front is passing overhead. Cumulonimbus clouds often mark the squall line of a cold front. "Often along the leading edge of the squall line is a low hanging arc of cloudiness called the shelf cloud" (NOAA n.d.). See photograph below.

> **The pressure will reverse direction within an hour when a front is passing overhead.**

Some visual features of Supercell Thunderstorms

A Supercell Thunderstorm has some visual features or aspects you can use to identify it:

1. Overshooting cloud

The anvil is a flat cloud at the top. A strong updraft will make an overshoot (bubble) of cloud on top of the anvil. If this bubble persists for over 10 minutes, it is a sign of a Supercell Thunderstorm.

Figure 14 (below): *Overshooting top cumulonimbus cloud.*

2. Rain-free base

Figure 15: *Rain shaft left, rain-free base of **supercell** thunderstorm right. Key West, Florida, U.S.*

3. Wall Cloud

If there is a separate cloud below both the rain-free base and the main updraft, it is called a wall cloud or pedestal cloud. It is a sign of a Supercell Thunderstorm. The wall cloud is located beneath the dark ***updraft***. The wall cloud can show streaks of cloud that suggest rotation.

Figure 16 (on following page): *Supercell Thunderstorm at Era Beach south of Sydney, Australia. The wall cloud (pedestal cloud) is lowered beneath the base of a cumulonimbus cloud. Picture also shows rain-free base that increases storm's longevity.*

Supercell Thunderstorms

The cores of supercell thunderstorms strongly rotate, and 70% of long-lived tornadic supercell thunderstorms produce tornadoes (Bunkers 2006).

A supercell thunderstorm is a Severe Thunderstorm with a persistent single, deep, rotating updraft or vortex. This rotating updraft, called a mesocyclone or 'medium-sized' cyclone, is found within a local-ised low-pressure area of a Supercell Thunderstorm and makes it a unique thunderstorm type. Supercell thunderstorms are recognised by their persistent rotating updrafts. Spiral bands and striations in the updraft tower are clues that the updraft is rotating.

Rotating supercell thunderstorms displace the updraft from the rain and possible hail in the downdraft, so they last much longer increasing the storm severity. It is a dynamic process resulting in a ***steady-state updraft***. Mature Supercell Thunderstorms move along almost straight-lines at nearly constant speed.

What is needed for supercells to develop?

What ingredients are needed for the development of supercells? Although supercells require some degree of buoyancy, moderate to strong speed and directional wind shear between the surface and about 20,000 ft (6,096 m) is the most critical factor. Wind shear not only creates the mesocyclone, but it also allows the storm to be tilted, which is important for maintaining a separate updraft and downdraft region. A separate updraft and downdraft allows the supercell to be long-lived because it reduces the likelihood that too much rain-cooled, stable air from the downdraft region will be ingested into the updraft, causing the storm to weaken (NOAA 2017).

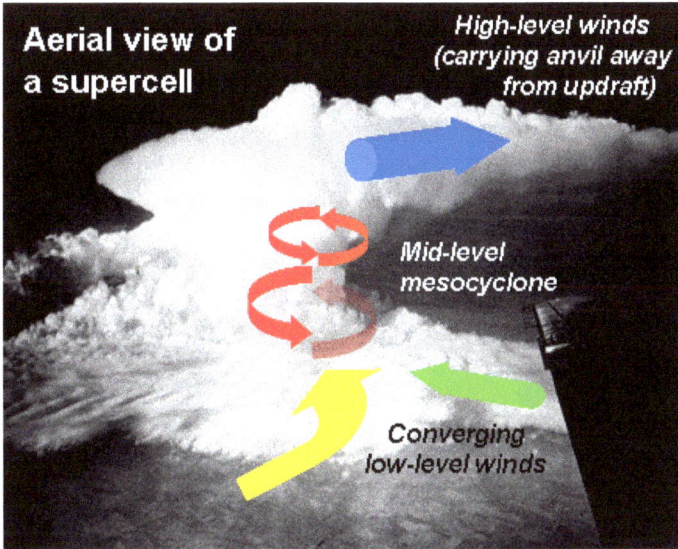

Figure 17: *Aerial view of a supercell thunderstorm. Photograph taken looking northeast over eastern Kansas, U.S.*

Supercell thunderstorms are a single rotating updraft thunderstorm large enough to be seen from space, but one or two orders of magnitude less in size than a cyclone. Their mesocyclone is a spinning vortex of air within the supercell thunderstorm. The rotation of the updraft allows it to remain separate from the downdraft(s), giving the storm greater longevity. Supercell thunderstorms are recognised in part by their sharp anvil cloud pointing downwind, an overshooting dome-shaped cloud and their rotation. A supercell thunderstorm is associated with the low-pressure region usually of a cold front. The barometer begins to fall steadily as the low-pressure region of a supercell thunderstorm approaches. The severity of the supercell thunderstorm means that the pressure starts falling at its maximum rate at the beginning of storm development up to 6 hours before storm onset.

Figure 18: *Intense updrafts produce a rain-free cloud base in a supercell.*

Supercell Thunderstorm Early Warning

A Supercell Thunderstorm Early Warning requires a:

- persistent rotating updraft (e.g. spiral bands and striations in the updraft tower)

- barometric pressure of 1005 mb or less

- first hour steady pressure fall from outset of more than 1.5 mb with an ongoing steady pressure trend.

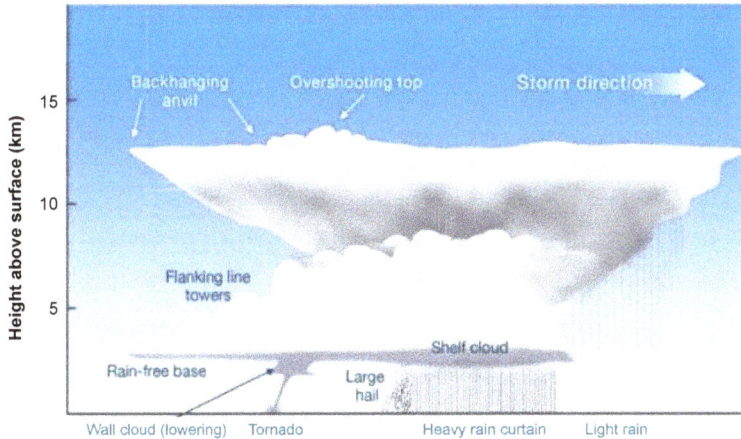

Figure 19: *Stylised supercell thunderstorm.*

Tornadic Supercell Thunderstorm Rule

A tornadic supercell thunderstorm requires all the following:

- persistent rotating updraft (e.g. spiral bands and striations in the updraft tower)
- barometric pressure of 1005 mb or less
- last 6 hours pressure fall at more than 1.5 mb per hour
- pressure curve has a 6-hour:
  - straight-line *trend* or
  - *rigid straight-line.*

A rigid straight-line will produce a deadly tornado (s).

Tornado Early Warning Rule

Tornado early warning requires all the following:

- persistent rotating updraft (e.g. spiral bands and striations in the updraft tower)
- barometric pressure of 1005 mb or less
- barograph displays the first hour steady pressure falls from storm outset greater than 1.5 mb.

- pressure curve has an ongoing:
  - straight-line trend, or
  - rigid straight-line.

A rigid straight-line will produce a deadly tornado (s).

You can now have *at least 5 hours early warning of tornado (especially a deadly one) from its signature on a barograph.*

A supercell thunderstorm that establishes a *rigid straight-line* steady pressure trend from the outset will tend to maintain it until the storm onset in about 6 hours. Good weather stations usually will have a barograph which will show a rigid straight-line trace with sufficient clarity.

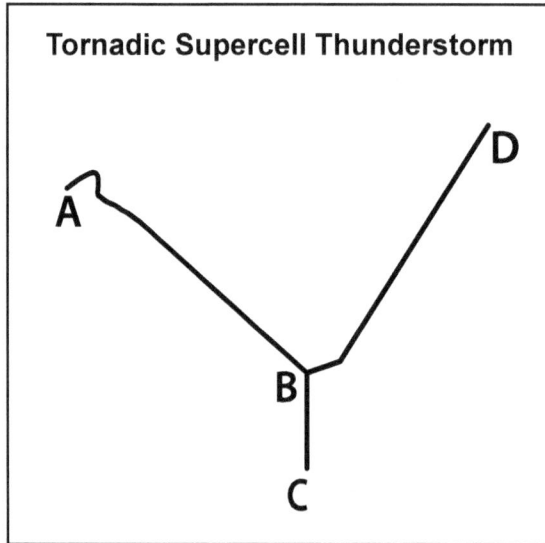

Figure 20: *Pressure profile – supercell with deadly tornado*

AB rigid straight-line on barograph means a *deadly tornado* will be produced.

BC tornado trace shows an immediate large sharp pressure spike and then an immediate large jump CB of similar magnitude.

BD shows the pressure curve rising from its minimum at B, remaining near constant for an hour or more before rising steadily to D.

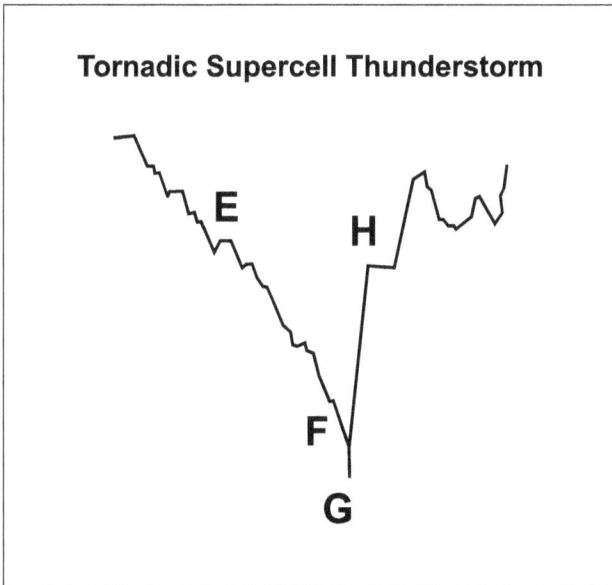

Figure 21: *Pressure profile – supercell with tornado*

EF straight-line trend on barograph means a *tornado* will be produced.

FG tornado trace shows an immediate small sharp pressure spike and then an immediate small jump GF of similar magnitude.

GH shows the pressure curve rising from its mimium at F steadily to H.

Supercell thunderstorms that do *not* have at least a 6-hour straight-line trend do *not* produce tornadoes. The trend is easy to see when the first 3 hours is a straight line.

The pressure in Figure 11 for the Severe Thunderstorm fell 13 mb in 7 hours to 1002 mb. It was not a supercell thunderstorm as its curve was very slightly concave and did not have the equivalent straight-line return rise of a supercell thunderstorm.

That the pressure curve has a straight-line trend for 3–6 hours until storm onset is the overriding consideration. You can get a tornadic supercell thunderstorm with a pressure curve that only falls 6 mb in 4.5 hours, but that is the exception. The strongest tornadic supercell thunderstorms will have a pressure curve with either a straight-line trend or rigid straight-line for 6 hours.

Strict adherence to the Tornadic Supercell Thunderstorm Rule predicts storms that have sufficient strength to produce a tornado. There are some supercell thunderstorms that do not produce tornadoes, but when you look at their pressure curve or profile they do not have the 6-hour straight-line trend or a rigid straight-line. It is the *rigidity* of the line that indicates **strength**.

A 6-hour straight-line trend with minor fluctuations will still produce a tornado. Observing that at least the *first* 3 hours of the 6-hour pressure fall is a straight line ensures that you have a clear straight-line trend while you are learning to recognise the trend.

A *tornadic* supercell thunderstorm will have 6-hour straight-line-trending barograph with only minor fluctuations, and can produce hailstones of baseball or cricket ball size or greater (7 cm or 2.76 inches diameter). Updraft speed will exceed 46 m/s (166 km/h or 103 mph). The stronger the updraft, the larger a hailstone can grow. The straight-line-trending barograph reflects the extremely strong rotating updraft that will produce a tornado.

The following are two 6-hour stylised barographs of tornadic supercell thunderstorms showing the trendline for pressure fluctuations trend or a rigid straight-line. The trendline is a line on the graphs showing the straight-line direction in which the curve appears to be going. A trendline goes through at least three peaks on the curve of a down trend.

Figure 22 (on following page): *Stylised barographs of supercell thunderstorm.*

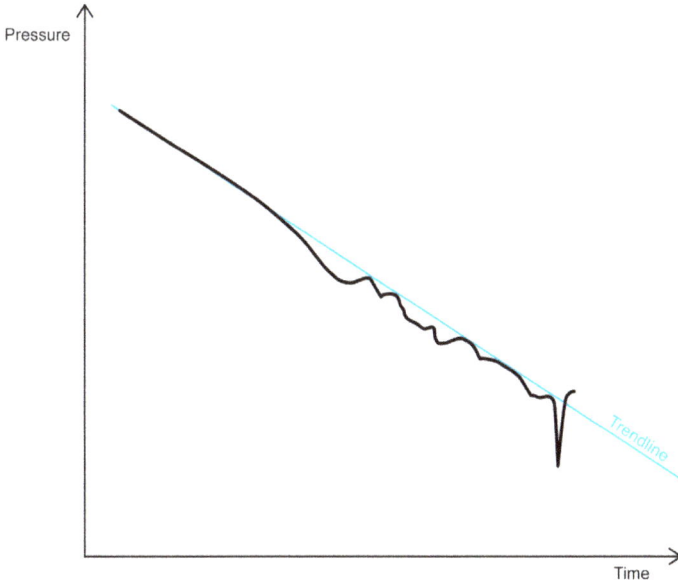

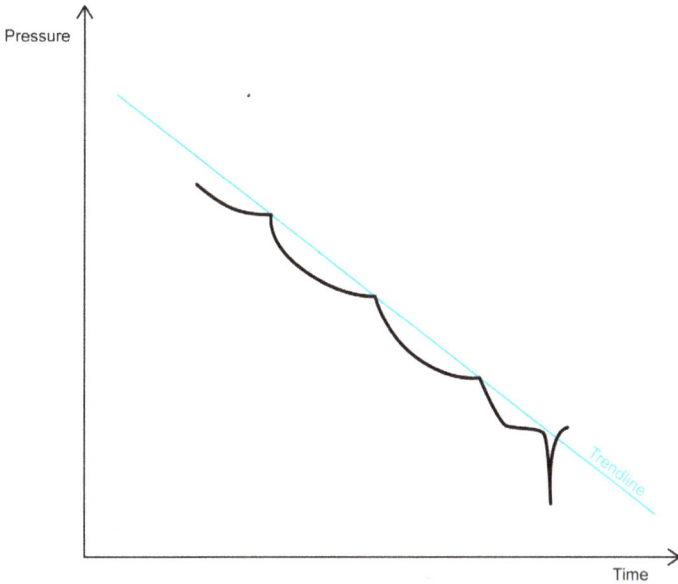

The barographs show a low pressure spike (tornado) at the end of a long period of steady falling pressure.

Tornadic supercell thunderstorm signature

The sides of the 'V' shape notch in the pressure curve are straight, signifying the storm's strength. This is the unmistakable supercell thunderstorm signature. In some cases when the curve reaches its minimum, the pressure can remain constant for an hour or more before rising steadily. Here, the pressure curve falls rapidly at a constant rate of 8 mb or more for 6 hours, flattens out and then rises rapidly at a constant rate for 6 hours or more.

Supercell thunderstorm updraft speed can reach 50 m/s (180 km/h or 112 mph) or higher. Typically, a non-supercell Severe Thunderstorm updraft speed is around 26 m/s (93 km/h or 58 mph) or higher. This difference in updraft speed marks a 'quantum' difference in updraft strength. The rotation of the updraft in a supercell thunderstorm causes the air to rise faster than that of a non-supercell thunderstorm, which is only driven by simple buoyancy of warm air rising.

It is important to distinguish between the barograph of a non-supercell Severe Thunderstorm and that of a supercell thunderstorm. The arrival of a non-supercell Severe Thunderstorm is preceded by a short sharp pressure rise and then some levelling off. The supercell in contrast has at least a 6 hours steady pressure fall in a straight line to a minimum. It will remain low and constant for an hour or more and then rise at a steady rapid rate for a least 6 hours.

Tornadoes

A tornado is "A violently rotating column of air touching the ground, usually attached to the base of a thunderstorm" (NOAA 2017).

More than 500 people worldwide can die in tornadoes in a year. The wind field of a tornado can be much larger than is visible to you. A larger tornado can contain or spawn other smaller but still deadly tornadoes. Winds flowing into a tornado can make it harder for you to escape.

A tornado is a windstorm. ***Understanding that it is a windstorm and not a thunderstorm is essential for its identification and prediction.*** A thunderstorm is most likely to spawn a tornado from a rotating wall cloud: a persistent wall cloud rotating rapidly and rising rapidly is likely to form a tornado. Most of these tornadoes would be categorised as intense tornadoes (FE3 or higher rating on the EF Scale) that are longer lived and stronger than the lower rated ones. The wall cloud, which may have an inflow of strong surface winds stronger than for non-tornadic storms, may immediately precede the formation of a tornado, although it usually persists for 10 to 20 minutes before the tornado appears. Tornadoes form inside a thunderstorm cloud and grow towards the ground.

Around 90% of tornadoes are weak (up to 180 km/h or 112 mph). 9% are strong and 1% are extremely strong.

Waterspouts form when cold air moves over warm water.

Figure 23: *Waterspout in the Gulf of Mexico photographed from the NOAA ship Rude. South of Cameron, Louisiana, Gulf of Mexico.*

Tornadoes often form on the boundary between the storm updraft and the downdraft. The updraft is observed as dark cloud where warm unstable air rises. The downdraft, usually white or clear, is where rain-cooled air sinks. The rear flank downdraft can be identified in a supercell thunderstorm as a 'clear slot' in the storm cloud, the bright clear sky you see at the back side of the 'dark' updraft. Tornadoes require strong vertical wind shear. They can last up to an hour, but 10 minutes is the average.

> A tornado is a violently rotating (usually counter clockwise in the northern hemisphere) column of air descending from a thunderstorm and in contact with the ground.
>
> The funnel cloud of a tornado consists of moist air. As the funnel descends the water vapor within it condenses into liquid droplets. The liquid droplets are identical to cloud droplets yet are not considered part of the cloud since they form within the funnel.
>
> The descending funnel is made visible because of the water droplets. The funnel takes on the color of the cloud droplets, which is white (NOAA 2016).

A barometer may start falling many hours or even days before a tornado if there is broad scale low pressure moving into the area. Steep pressure falls often happen as the parent or main circulation in the thunderstorm moves overhead or nearby. The largest pressure fall will be in the tornado itself.

You need to identify:

- a cloud base that is rapidly rotating and
- whirling debris or dust cloud near the ground under the thunderstorm base (Edwards 2010).

Supercell thunderstorms can spawn tornadoes at the back of the storm under the dark rotating updraft. The inflowing updraft will blow into your back if you stand facing a storm. A rotating updraft will be present at least 20 minutes before a tornado is spawned.

Figure 24: *Twin violent (EF4) tornadoes, Wisner, Nebraska, U.S.*

Other tornado early warning

Jump in lightning flash rate

Lightning flash rates depend on the storm updraft speed. The higher the updraft speed, the higher the flash rates overall (Del Genio 2007). When tornado development begins, the updraft of the parent thunderstorm is augmented causing the thunderstorm flash rate to *jump,* giving a tornado harbinger.

A lightning jump is usually a sign of strengthening updraft. Lightning jumps have occurred before severe hail and a tornado touchdown (Schultz 2014). You need to watch the lightning rate once it jumps to 10 flashes or more per minute (Schultz 2009).

Small surge in updraft

There will be a sign in the parent supercell thunderstorm's *pressure trace* (small pressure fall) before severe hail and a tornado touchdown corresponding to the small surge in updraft speed. The barograph signature of an incoming supercell thunderstorm (with severe hail) shows a characteristic pressure dip, then after a short rise, another dip for the hail release and finally, the usual short steep rise in pressure. This second dip is a pressure response to the small surge in updraft speed. *A small dip in the parent supercell thunderstorm's pressure trace is expected for a tornado touch down.* This is a tornado early warning.

Online monitoring of a tornado

As flying debris is the biggest concern with tornadoes, you can safely view and closely track the progress of a tornado, without putting yourself in harm's way, by using an online weather station such as www.wunderground.com. Globally, there are over 250,000 weather stations, some staffed and others automated. The secret is to find a weather station close to where you need to track the tornado, although occasionally, a weather station too close to one will be damaged. Weather stations usually have a barograph showing a trace of the barometric pressure over time.

Summary: online monitoring of a tornado

1. Choose the place where you wish to monitor the pressure.
2. Find the nearest online weather station to that location.
3. Track the pressure and its rate of change to detect a tornado in the area and predict its severity.
4. As the tornado comes through, observe its features to confirm its severity.

Tornado signature

The barograph signature of a tornado is different from that of an incoming thunderstorm, which shows a characteristic pressure dip with a short steep rise in pressure immediately after the pressure dip.

The barograph signature of a tornado shows an immediate large sharp pressure fall and then an immediate large jump of similar magnitude.

The Tornado Signature tells you that a tornado is present even though it may not have a visible funnel.

Signature shape shows tornado proximity

The shape of the tornado signature changes depending on how far away the passing tornado is from the barograph recorder. The greater the distance between you and the passing tornado, the shallower and wider the pressure dip will appear.

This is illustrated in the following barographs.

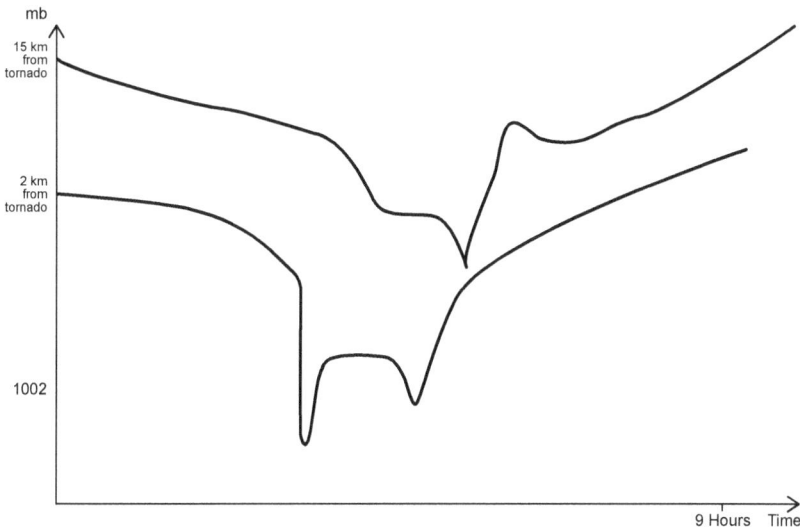

Figure 25: *Signature shape shows tornado proximity.*

Non-Supercell tornadoes

Supercell thunderstorms are recognised by their persistent rotating updrafts. Spiral bands and striations in the updraft tower are clues that the updraft is rotating. Eastern Colorado receives non-supercell tornadoes but they are generally small and they occur in remote areas.

The largest tornadic supercell thunderstorm

The largest tornadic supercell thunderstorm's pressure dropped over 12 mb in 6 hours before onset and spawned a tornado. The sides of the 'V' shape pressure curve were straight, signifying the storm's strength. It produced the legendary long-track Tri-State tornado of March 1925 (Maddox et al 2013). A non-supercell thunderstorm curve has a slightly concave shape. The tornado sharp vertical pressure falls when the pressure due to the supercell thunderstorm reaches its minimum pressure.

When the lightning region instantly disappears

A supercell thunderstorm that establishes a strong steady pressure trend from the outset will tend to maintain it until the storm onset in about 6 hours. Lightning from the thunderstorm would otherwise occur with these conditions for non-supercell or smaller storms. No lightning will be observed while the storm continues to develop. If the pressure curve is slightly concave and downwards, it will be a thunderstorm or Severe Thunderstorm, *not* a supercell.

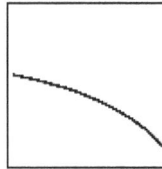

Figure 26: *Concave down curve.*

If there is lightning in a cloud below a wall cloud that has formed since storm onset, you have a supercell thunderstorm only, not a supercell thunderstorm and a tornado. A tornado vortex cloud does not produce lightning, whereas a thunderstorm does. So, if there is *no lightning* in rotating cloud *below* the wall cloud, it is a tornado. There will still be lightning coming from the thunderstorm but not from a tornado, so you will need to observe carefully. *There is an even earlier warning sign when the lightning region in the updraft of the supercell thunderstorm instantly disappears. Seconds later, a tornado will emerge.*

The Tornado Early Warning Rule can sometimes give you as much as 6 hours early warning, when the supercell thunderstorm remains in a near steady state for 3 hours before the rear flank downdraft descends to the ground spawning a tornado.

Tornado rule for non-supercell thunderstorms

In east England where non-supercell tornadoes can occur, the pressure can fall over 8 hours, by 8 mb or more, to below 1005 mb, and then spike down. If the pressure curve *begins to turn concave down by* 0.25 mb in 30 minutes, *and the downturn accelerates, this indicates a tornado will occur within an hour*. In general, if the pressure falls 4 mb (or more) in 15 minutes or less, it is a tornado.

It is vital to closely monitor the curve from the time the 6 mb fall is reached.

Example

The 'Eagle Tornado' struck in the Eagle area, Waukesha County near Milwaukee US on 21 June 2010. The pressure fell steadily from 1018 mb to 1010 mb in 11 hours but then it fell sharply in 15 minutes before striking, which is characteristic of a tornado. The pressure immediately jumped sharply after the sharp fall, in this case, over 8 mb.

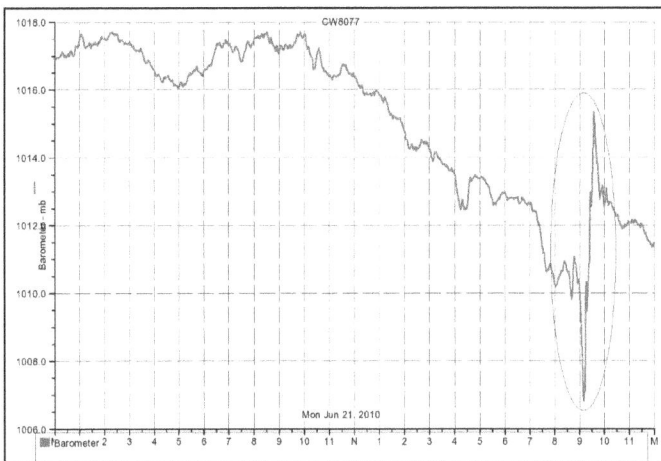

Figure 27: *Barograph of a tornado (Jensen 2010).*

48

The pressure signature above is typical of areas where tornadoes are prevalent, such as St. Louis, Missouri US.

Tornadoes: summary of other warning signs

The following signs give you warning of a tornado.
- Pressure fall of 4 mb or more in 15 minutes or less.
- Lightning flash rate jump.
- Tornadic supercell thunderstorm has formed.
- Strong vertical wind shear.
- Distinct boundary between the storm updraft and the downdraft.
- Rotating wall cloud persisting for 10 minutes or more.
- Whirling debris near the ground under the thunderstorm base.

Hailstorms

Hail damage from increasingly furious storms is a leading cause of insured losses. People and livestock can die in hailstorms. Damaging hail is produced by Severe Thunderstorms.

Prediction of large hail is best handled using a good barograph app with pressure sensor and the simple rules for predicting Severe Thunderstorms presented in this book. See page 27 for the *Severe Thunderstorm Early Warning Rule*. Also, see *Supercell Thunderstorm Early Warning*, page 35. Ordinary people can now have time to prepare for hailstorms keeping them safe.

The increasing intensity and lifespan of tropical cyclones will necessitate a new category 6 of storm on the Saffir-Simpson Hurricane Wind Scale. It would better accommodate the 320km/h (200 mph) storms we have seen globally in recent years and increase awareness that global warming is making the strongest storms even stronger.

The author 'predicts there will soon be a need for a new class of hurricane categorisation. He believes the rise in ferocity and destructive capacity of storms fuelled by climate change is such' *(McGuire, 2019)*.

Figure 28 (on following page): *Lightning shoots up updraft and anvil of tornadic supercell at night, with car light trails.*

Cyclones/Typhoons/Hurricanes

Tropical revolving storms are referred to as 'cyclones' in the Indian Ocean and South Pacific, as 'typhoons' in the Western Pacific, and as 'hurricanes' in the Western Atlantic and Eastern Pacific Oceans. A tropical revolving storm satisfies the barometer rule for a Severe Thunderstorm and so does a tropical storm.

The Thunderstorm Heat Engine Cycle is shown below.

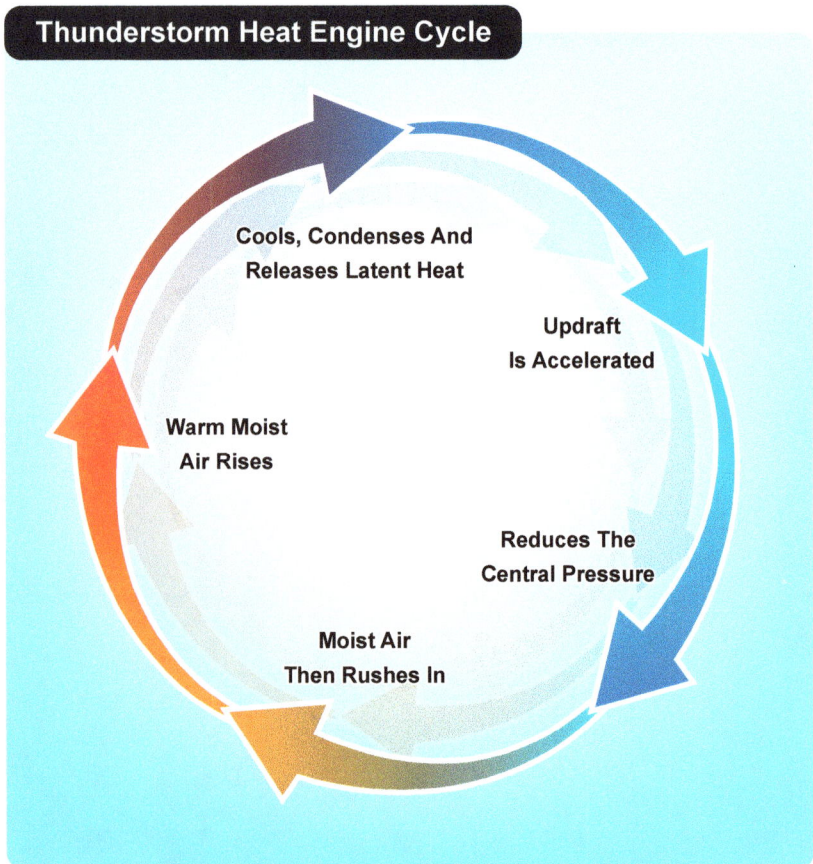

Thunderstorm Heat Engine Cycle

Cools, Condenses And
Releases Latent Heat

Updraft
Is Accelerated

Warm Moist
Air Rises

Reduces The
Central Pressure

Moist Air
Then Rushes In

Figure 29: *Thunderstorm heat engine cycle.*

The hurricane has an eye. The eye is almost calm. It is the warmest region at the centre of the hurricane and it has the lowest pressure. It is surrounded by an eye wall having the thickest cloud, heavy rain and strongest wind.

The **Hurricane Heat Engine Cycle** is similar to that of the thunderstorm at the start. Warm air rises and condenses when it cools. The rising air reaches the top of the troposphere and spreads out in all directions. It differs from the thunderstorm as some air does not escape as it spreads out but comes back down the eye. This warm air is trapped by the rotating vortex of wind. The warm moist air compresses inside the eye and this compression heats the eye further. The central pressure in the eye keeps dropping. The process becomes self-sustaining as warm moist air over the warm ocean rushes in.

A cyclone/typhoon/hurricane requires all the following:

- barometric pressure of 990 mb or less
- last 3 hours pressure fall of 4 mb or more
- last 8 hours pressure fall of 8 mb or more.

This rule can be used for a cyclone/typhoon/hurricane approaching landfall.

Steady pressure in tropics is hurricane harbinger

Cyclones/hurricanes/typhoons occur in the tropics in late summer and autumn. There is a twice daily rise and fall of pressure on Earth, which is more marked in the tropics. It peaks around 10 am and 10 pm. In the absence of a hurricane, the barometric pressure in the tropics varies about 3 mb each day. At least 30 hours before an approaching hurricane making landfall, the pressure becomes almost steady for 6 hours.

This **6-hour steady pressure interval** that occurs before the cyclone's pressure dip is an important early warning sign of a hurricane's development. It is the **Hurricane Harbinger**.

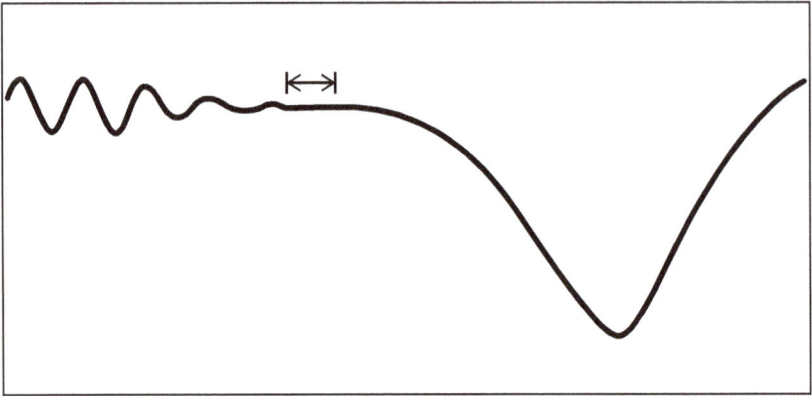

Figure 30: *Stylised barogram showing 6-hour steady pressure interval.*

The steady pressure interval, after which the pressure dips, is unique for each cyclone and its value is near the seasonal average pressure (e.g. 1016 mb or 30 in). The 6-hour steady pressure interval for a Category 3 cyclone (Category 2 hurricane) commences at least 30 hours before landfall. See Tropical Cyclone Classification at: http://www.srh.noaa.gov/jetstream/tropics/tc_classification.html).

Even if the pressure curve is not entirely flat for the last 6 hours, significant flattening will still be apparent from the barograph, giving you early warning.

Hurricane Early Warning Rule

24-hours early warning of a hurricane is when the pressure has been almost steady for 6 hours.

Remember that the Hurricane Early Warning Rule also applies to cyclones and typhoons.

Figure 31: *Hurricane viewed from satellite.*

Firestorms

Wind is air pressure on the move. When air pressure changes rapidly, it can produce a windstorm. A hot, dry windstorm and fire can produce a firestorm. A firestorm (short for fire thunderstorm) forms when heat from a wildfire or bushfire creates its own wind system. The updraft of a rising column of hot, dry air over a large intense original fire mushrooms and causes *low and rapidly falling barometric pressure*. The pressure gradient causes Gale Force winds to converge on the fire, bringing in more and more cooler surrounding air to the lower-pressure centre. This is called the 'chimney effect'.

Clouds from bushfires are grey or brown but the pyrocumulonimbus clouds that rise above the smoke from bush firestorms are white. 'Dry' lightning is produced in a firestorm as the threshold updraft speed of 7 m/s (25 km/h) for lightning is exceeded. Lightning can potentially spread the firestorm. Climate Change is significantly increasing the updraft speed and lightning flash rates of bush firestorms.

PREDICTING FIRESTORMS

Bushfires or wildfires can produce firestorms if the conditions are suitable.

Your barometer and thermometer are powerful instruments for predicting firestorm conditions.

Firestorm Conditions

Firestorm conditions occur when a fire exists and a pressure fall of 10 mb or more has occurred during the last 4 hours provided that:

- low pressure is sustained below 1000 mb,

- the temperature is sustained above 40°C, and

- the relative humidity is 10% or less

Firestorm Early Warning

Early warning of firestorm conditions is when a fire exists and the **pressure starts falling at over 1.5 mb / hour** while

- the temperature is sustained above 40°C,

- low pressure is sustained below 1000 mb, and

- the relative humidity is 10% or less

The relative humidity before the onset of a fire thunderstorm will plunge to 10% or less whereas the relative humidity for a normal thunderstorm needs to be above 80%.

You need to use your barometer in close, but safe, proximity to the fire. You may be able to remotely monitor the pressure at a weather station near the fire using the internet. Firestorms have been known to have the energy equivalent of up to 40 times that of a typical wildfire. The rapid fall in barometric pressure before the onset of a firestorm can cause impaired concentration and relatively poorer decision-making. You cannot fight a firestorm. Immediate evacuation is your only option.

The risk of firestorms has increased as the number of hot days and the length of heatwaves have increased through human-induced Climate Change. Firestorm events appear to be increasing worldwide in number and strength. They are occurring in places where they have never been known to occur previously. A firestorm in California in 2018 generated a tornado-strength vortex. Hurricane Force winds gusting to 160 km (100 mph) were generated by a firestorm in California in 2019.

Just as a supercell thunderstorm can spawn a tornado so a fire thunderstorm can spawn a fire tornado. The world's first fire tornado was recorded in a firestorm in 2003 in Canberra, Australia. The emergence of this new phenomenon marked a tipping point in global warming. Fire tornadoes are tornadoes rating at least an EF2 (181 to 253 km/h) on the Enhanced Fujita Scale that classifies tornadoes by their estimated wind speed. On 31 December 2019, a fire tornado overturned a 10-tonne truck at Jingellic, NSW in Australia killing the occupant.

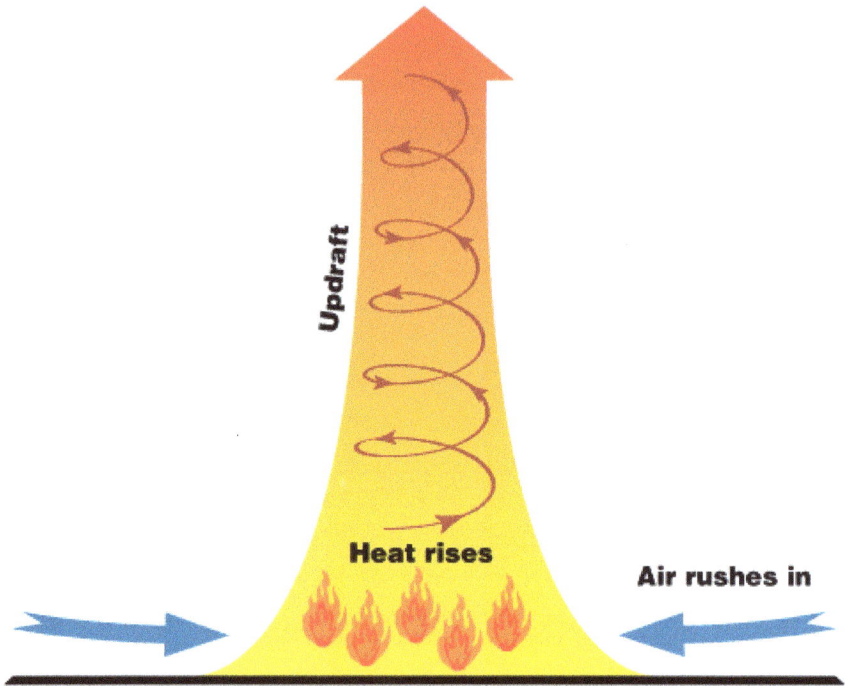

Figure 32: *How firestorms form.*

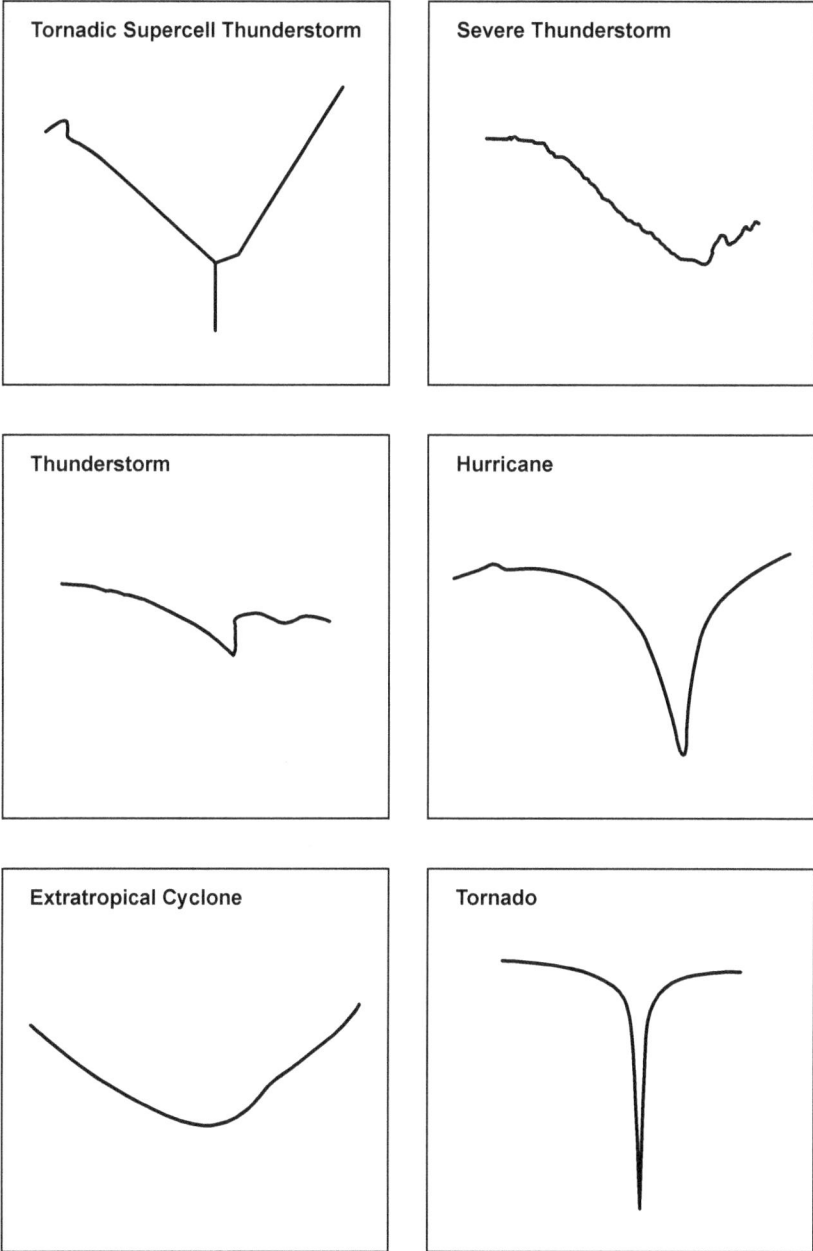

Figure 33: *Stylised barographs of typical storm signatures.*

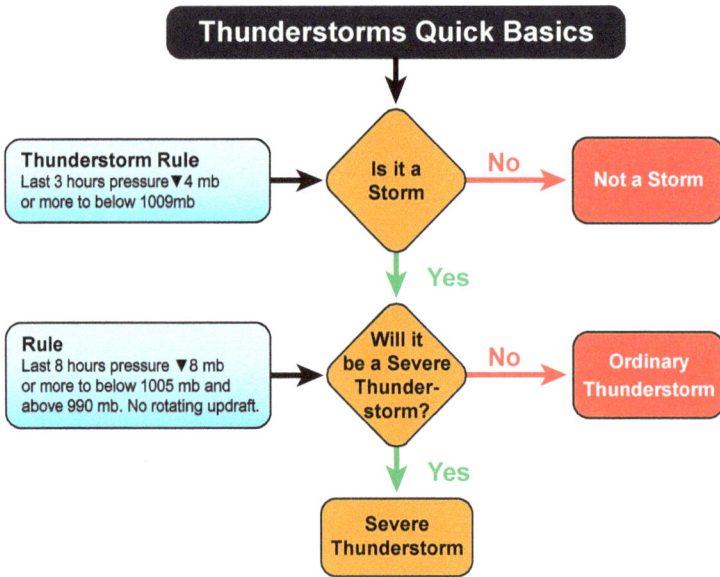

Thunderstorms Quick Basics

Thunderstorm Rule
Last 3 hours pressure ▼4 mb or more to below 1009mb

Is it a Storm — No → **Not a Storm**

Yes

Rule
Last 8 hours pressure ▼8 mb or more to below 1005 mb and above 990 mb. No rotating updraft.

Will it be a Severe Thunder-storm? — No → **Ordinary Thunderstorm**

Yes

Severe Thunderstorm

Thunderstorm

If the Rule is satisfied the storm will begin when the short steep rise or jump in pressure is completed after the pressure dip. The temperature makes a corresponding short steep fall when the storm begins.

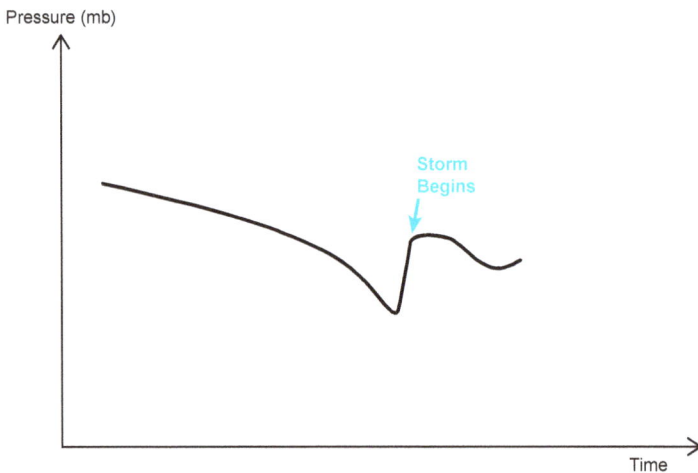

Pressure (mb)

Storm Begins

Time

Figure 34: *Thunderstorm quick basics.*

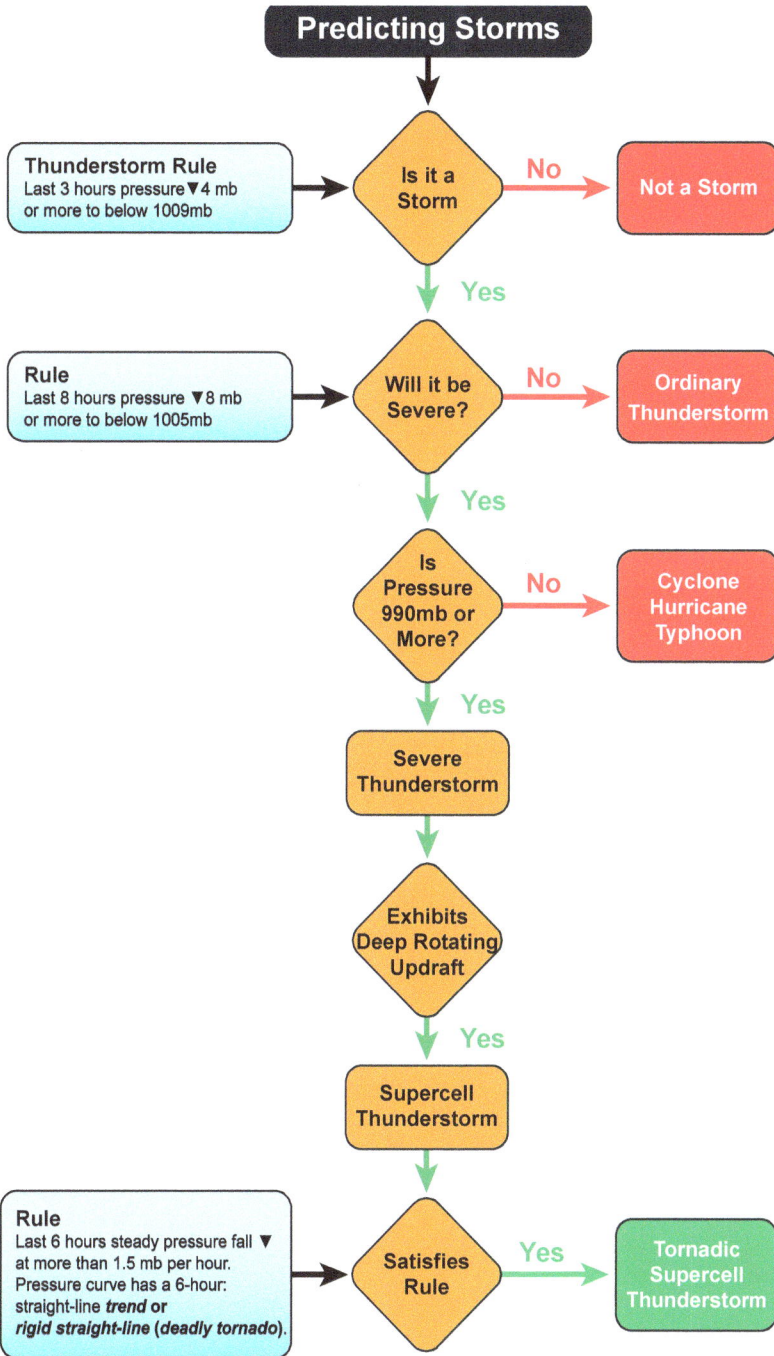

Predicting Storms

Thunderstorm Rule
Last 3 hours pressure ▼4 mb
or more to below 1009mb

Is it a Storm — No → **Not a Storm**

Yes

Rule
Last 8 hours pressure ▼8 mb
or more to below 1005mb

Will it be Severe? — No → **Ordinary Thunderstorm**

Yes

Is Pressure 990mb or More? — No → **Cyclone Hurricane Typhoon**

Yes

Severe Thunderstorm

Exhibits Deep Rotating Updraft

Yes

Supercell Thunderstorm

Rule
Last 6 hours steady pressure fall ▼
at more than 1.5 mb per hour.
Pressure curve has a 6-hour:
straight-line *trend* or
rigid straight-line (*deadly tornado*).

Satisfies Rule — Yes → **Tornadic Supercell Thunderstorm**

Figure 35: *Predicting storms.*

60

Barometer Rules

Last 3 Hours

▲10 mb	Gale warning (34–47 knots)
▲6 mb	Strong wind warning (26–33 knots)
▼4 mb	Storm (< 1009 mb)
▼6 mb	Storm with strong wind warning (< 1009 mb)
▼10 mb	Firestorm conditions if fire exists and (< 1000 mb) & (>40°C) & (Relative Humidity < 10%) sustained (4 Hours)
▼10 mb	Storm with a gale warning (< 1009 mb)

Severe Thunderstorm Warning

Last 3 hours ▼4 mb or more

Last 12 hours ▼8 mb or more

(Pressure < 1005 mb)

Tornado Early Warning Rule

Persistant rotating updraft (e.g. spiral bands/striations in updraft tower)

Barometric pressure of 1005 mb or less

Barograph: first hour steady pressure falls from storm outset greater than 1.5 mb.

Pressure curve has an ongoing straight-line *trend, or rigid straight-line*.

A *rigid straight-line* will produce a *deadly tornado(s)*.

At least 5 hours warning of tornado from its signature on a barograph.

Cyclone / Hurricane / Typhoon Warning

Last 3 hours ▼4 mb or more

Last 8 hours ▼8 mb or more

(Pressure < 990 mb)

Storm Season (Tropics)

Steady	Hurricane early warning if pressure almost steady for 6 hours

Figure 36: *Barometer rules.*

2

Sea and Surf

Warm moist air is the fuel of storms. If a storm moves from land over a stretch of warm ocean, it can strengthen rapidly. Storms are more frequent over the sea. Hurricanes weaken quickly once they make landfall.

The barometer rules already given apply to storms on land or at sea. Below are useful methods to confirm your barometer predictions made at sea.

Early Warning of Storms

Swell waves are waves remotely generated from a storm or wind. They propagate in wave trains with more regular and longer periods and are of similar height. They can travel hundreds of kilometres away from the storm that produced them. Their direction of travel can be different from that of the original prevailing wind. A swell does not need the prevailing winds to sustain it.

Swell height is the wave peak-to-trough size. The wave period is the time in seconds that consecutive wave crests take to pass a point.

> The wave period is the time in seconds that consecutive wave crests take to pass a point.

A forecast is given as height@period. For example, a surf forecast may be: *Swell tomorrow will be 1.8 m@10 seconds.*

Sea waves are generated by local winds. Sea waves exhibit a jumbled collection of periods and heights.

A swell from a distant storm has a period of 10 seconds or more. A period of 25 seconds was recorded from a storm in the Southern Ocean. Knowing the swell period is essential to making accurate surf forecasts and calculating the time before swell waves arrive from a distant storm.

The standard dominant wave period (DPD) from buoy data taken on 9 November 2016 (NOAA 2017) showed a 'DPD of 6 seconds' (10 crests arriving per minute). This is typical of normal swell without the influence of a distant storm. After a distant storm, the arrival rate *slows noticeably*—it almost halves from 10 to 6 crests per minute, that is, the fastest waves (longest period) are first to arrive and will have a period of at least 10 seconds (arriving at only 6 or fewer crests per second). This slowing is an early warning sign or harbinger of an approaching storm (NOAA 2017).

Figure 37 (on following page)*: Swell lines in the Pacific.*

A longer swell period gives a severe storm more energy and the swell travels faster.

On 20 August 2015, Hurricane Kilo, a large and intense storm, had a swell of 2.5 m@18 seconds. A smaller more common hurricane at Santa Rosa Island in August 1932 had a swell period of 10 seconds.

Normally without a hurricane, wave crests can arrive at a rate of around 10 per minute. However, wave crests from a distant hurricane can arrive much more slowly, at a rate of around 6 per minute, and this is quite noticeable. The number of crests passing a point per minute almost halves (10 per minute to 6 per minute) when swell waves arrive from a distant storm. This provides an early warning of the approach of a hurricane.

All storms are unique and the examples discussed are estimates. The swell generated by a larger hurricane can have 4 crests passing a point per minute which has a swell period of 25 seconds. Swells with this period would be among the first to arrive from a distant hurricane.

Swell waves travel in groups—called sets by surfers—with the tallest in the middle. In a group of 14 waves, the seventh would be the tallest. A group of 7 or more waves can come from a local storm, whereas a group from a distant storm can be 3 times larger.

Surfers hope for large swell height with a long period. As you move away from the source of the storm, the swell gets better for surfing but the swell height decreases. The challenge for the surfer is to find the best balance in this trade-off.

Secret of the Swell

Dissipation (weakening) is stronger for waves with a shorter period. Dissipation of swell waves from a distant storm where the period is 13 seconds or greater is so weak that the swell waves will be sustained when they arrive many thousands of kilometres away from the storm. These swells bring high-quality surf, the 'secret of the swell' for surfers.

Most surfers refer to waves where the period is 13 seconds or greater as 'ground swell' and information on the wave period is available from buoys. Waves with a period of 10 to 12 seconds can still be worthwhile for surfers. So any swell with 6 or fewer wave crests passing a point per minute (10-second period or more) is suitable for surfers and serves as a harbinger of a distant storm. However, the *secret of the swell is to have a period of 13 seconds or greater.*

The most suitable swell for surfers will have a narrow range of long periods.

You can hear the rhythmic beat of the swell as it laps against the shore or a boat. If the beat slows to 6 or fewer per minute, there is a distant storm and strong winds to come. You can calculate wave period at sea using a stopwatch. Measure the average crest-to-crest time for a floating object to move through a number of waves. If the period is 10 seconds or more, a distant storm and strong winds are to come.

When swell waves arrive at Newport Beach, California from a distant storm in the North Pacific, the longest period waves—the fastest—arrive first, followed by the next longest and so on.

Figure 38 (below): *Swell from a distant storm apparent on the sea surface.*

An example is the Florida Keys Labour Day Hurricane that made landfall at Cedar Key on 4 September 1935. This was a Category 2 Hurricane. The table below shows for a typical Category 2 hurricane / Category 3 cyclone approaching a coastal area that from 96 hours before landfall the swell had a period of 10 seconds for the next 24 hours, 9 seconds for the next 24 hours, 8 seconds for the next 12 hours, and so on.

Hours Before Landfall	Swell Period (seconds)	Hours with each Period
96	10	24
72	9	24
48	8	12
36	7	4
30	5	6

Table 3: *Typical Category 2 hurricane/Category 3 cyclone approaching coastal area.*

To calculate the speed in knots of a swell, you multiply the spell period by 1.5. For example, a swell with a 10-second period travels at 15 knots.

How long for waves from distant storms to reach my beach?

If you want to know how long it will be before waves from a distant storm reach your beach, you divide that distance by the swell speed. "So if a storm that is 2,100 nautical miles away generates swell with a period of 14 seconds, that surf will show up at your shore in 2,100nm/(14 x 1.5) knots = 100 hours or 4 days and 4 hours" (Borg 2015).

Figure 39: *Swell.*

Hours Before Landfall	Swell Height	Swell Period (seconds)
96	1 m (3 ft)	10
72	2 m (6 ft)	9
48	3 m (9 ft)	8
36	4 m (13 ft)	7
30	–	5
24 (Evacuation completed by this time)		

Table 4: *Hours before waves make landfall.*

Estimating Hours before a Hurricane makes Landfall

The example in Table 4 gives the general sequence of events for a Category 2 hurricane / Category 3 cyclone approaching a coastal area. You can work out the hours left before landfall by observing the swell period shown below. For example, a swell period of 7 seconds indicates 36 hours are left before landfall.

Estimating the Wind

The sea state describes the effect of local winds on conditions. You can use the sea state to estimate the wind.

Force	Knots	At Sea
0	0	No waves—glassy
1–3	1–10	Small smooth wavelets
4	11–16	Small waves—occasional white horses
5	17–21	Persistent white horses
6	22–27	Large waves begin to form, more foam crests, some spray
7	28–33	Waves heap, streaks along wind direction
8	34–40	Gale, waves 5.5 m, streaks prominent
9	41–47	High waves and dense streaks of foam
10	48–55	Very high waves and spray makes sea almost white
11	56–63	11-m-high waves and wave crests to froth
12	64 +	Air filled with foam and spray, visibility low

Table 5: *Estimating the wind—at a glance.*

Where you are in Relation to Low or High-Pressure Systems

The wind direction can help in determining where you are in relation to a low or high-pressure system. Here is a useful rule:

In the Northern Hemisphere:

Stand with your back to the wind.

Turn 30° to your right.

Low pressure will be on your left and high pressure on your right.

In the Southern Hemisphere:

Stand with your back to the wind.

Turn 30° to your left.

Low pressure will be on your right and high pressure on your left.

Crossed Winds Rule

The Crossed Winds Rule (Watts 2014) tells you the direction from which a change will come. This rule is applied when the weather is deteriorating and the direction of the cirrus clouds crosses the direction of movement of the surface winds of a frontal system. The slight 30° rotation is an added improvement.

Northern Hemisphere

Stand with your back to the wind.

Turn 30° to your right.

If the cirrus clouds are coming from your left,

that's where the low pressure is located,

and from which change will come.

Southern Hemisphere

Stand with your back to the wind.

Turn 30° to your left.

If the cirrus clouds are coming from your right,

that's where the low pressure is located,
and from which change will come.

Gusts

Gusts at sea can be dangerous, as the wind force increases by the square of the wind speed. Downdrafts carrying the high wind speed of upper winds can produce gusts. Downdrafts can occur from a cumulus cloud with a cold front associated with a squall or squall line. If you see Virga hanging from clouds in this situation, expect strong downdraft gusts.

Virga clouds look like wispy veils of precipitation that evaporates away. They look like jellyfish tentacles floating in the sky.

Figure 40 (below): *Virga viewed SW from Flat Top Mountain, North Carolina, U.S.*

3

Predicting Rain

Barometer Rules for Rain	
Last 3 Hours	
0.1–1.5 mb ▼	Rain (< 1009 mb)
4 mb ▼	Rain (1009–1023 mb)
4 mb ▼	Rain to come (> 1023 mb)
1.1–2.7 mb ▲ Humidity > 90%	Rain (> 1013 mb)

Table 6: *Barometer rules for rain.*

The amount of rainfall produced by a thunderstorm depends on the air moisture content, wind strength and wind shear. Strong low-level winds can produce thunderstorms with moisture. Low wind shear encourages heavy rainfall and increased rainfall efficiency. The presence of weather fronts, a low-pressure system and topography also influence the amount of rainfall produced by a thunderstorm.

An Altimeter Can Indicate Stormy Weather

This section will be of interest to pilots, travellers, hikers, campers or anyone who uses an altimeter. An altimeter measures atmospheric pressure and calculates altitude. You can also use an altimeter to measure a change in barometric pressure at a particular altitude. Most popular smartphones have an altimeter. If you remain at the same altitude, the altimeter will show an apparent increase in altitude with falling barometric pressure. An apparent fall in altitude of 36.6 m (120 ft) or more in the last 3 hours indicates a storm, or rain or rain to come.

An apparent fall in altitude on your altimeter of 36.6 m (120 ft) or more in the last 3 hours indicates a storm, or rain or rain to come.

Rapid Pressure Fall Foretells Storm or Rain

If the last 3 hours ▼ 4 mb or more:

pressure > 1023 mb	rain to come
1009–1023 mb	rain
pressure < 1009 mb	storm

4

Cloud Sequences

Sequence of Cloud Arrival for Storms

Cloud Types		
	Height	**Name**
High	Greater than 6,000 m	**Cirrus cirrostratus cirrocumulus**
Middle	2,000–6,000 m	**Altostratus altocumulus**
Low	Less than 2,000 m	**Stratus stratocumulus Nimbostratus Cumulus cumulonimbus***

Table 7: *Cloud sequences.*

*occupies low to middle altitudes

Figure 41: *Cirrus at sea.*

Cirrus clouds indicate the wind direction of jet stream level winds, which are narrow bands of strong wind in the upper atmosphere. When you can see only cirrus clouds, expect fair weather for the next 12–24 hours. The further things are away from you, the slower they appear to move. You would expect very distant cirrus clouds to appear almost stationary. So, if you see them moving, they must be going very fast and you can expect a strong change. Subsequent clouds arriving after the cirrus on the move will be thicker and at progressively lower levels in the atmosphere. Cirrus clouds signal the beginning of the sequence. They can indicate that moist air from a distant storm has risen into the upper atmospheric layers. You can tell the direction from which a front is approaching by watching the cirrus clouds. Cirrus clouds are seen with fronts, thunderstorms and cyclones/hurricanes/typhoons.

When cirrostratus immediately follow cirrus clouds, you can usually expect a storm or a snowstorm in 12–24 hours.

Figure 42 (below): *Cirrocumulus cloud, Michigan, Grand Rapids, U.S.*

Cirrus, cirrostratus and altostratus (uniform grey sky) signal the approach of a *warm front*. Stratus and nimbostratus complete the sequence for the warm front. However, if cirrocumulus clouds—called 'mackerel sky'—immediately follow the cirrostratus ('mares' tails'), it signals that a *cold front* with gusty strong winds and rain is approaching. If the warm air associated with the cold front is unstable, the cloud sequence will be cirrus, cirrostratus, cirrocumulus, altocumulus, cumulus, cumulonimbus, stratocumulus. Altostratus and nimbostratus clouds can occur at a cold front if the warm air region of the front is stable.

Cold fronts have taller cumulus clouds under the higher altitude cloud types than warm fronts. Cumulonimbus clouds producing thunderstorms and squalls often form along a front.

Cirrus clouds are seen at the top of thunderstorms (Lydolph p.122). Cirrus arrive first from a distant thunderstorm, followed by cirrostratus and cumulonimbus.

Figure 43 (on following page): *Sun halo and cirrostratus clouds at sunset.*

From a cyclone/hurricane/typhoon, cirrus clouds approach 36 hours before landfall.

High level

The presence of cirrus clouds, which at < 6,000 m (20,000 ft) are very high, means the fair weather will change in a day or two. They indicate that moist air has risen into the upper atmospheric layers. Cirrus clouds contain ice crystals. Light from the moon or sun can be refracted through these hexagonal ice crystals, appearing to give a halo effect.

Cirrostratus are high-level relatively transparent sheets of ice crystals.

Middle level

Altostratus clouds (2,000–6,000 m; 6,500–20,000 ft) are blue-grey and cover the sky.

Figure 44 (below): *Mackerel sky of altocumulus clouds over eliptic crater, Erta Ale volcano.*

Altocumulus (another type of 'mackerel sky') signals convection and middle level atmospheric instability. Pilots are cautious when encountering towering altocumulus castellanus cloud with its castle turret-like protrusions above the main cloud deck. They may be accompanied by turbulence or even icing. If you see this cloud in the morning, expect a thunderstorm in the afternoon. You can confirm this using the Thunderstorm Rule.

Low level (< 2,000 m; 6,500 ft)

Different clouds bringing different weather can arrive as follows:

- Stratus clouds are a uniform grey colour and can bring some drizzle
- Nimbostratus clouds can bring rain or snow. Nimbostratus clouds are typically produced by a warm front. They can occur ahead of the front, bringing warmer temperatures. Fog can also occur ahead of a warm front
- Cumulus clouds (fair weather types)
- Cumulonimbus clouds are associated with a cold front and can bring thunderstorms. Towering cumulonimbus clouds can grow to 6000 m (20,000 ft)
- Stratocumulus clouds are low, lumpy grey or dark grey clouds produced by strong cold winds. They are usually the last clouds to arrive before a front.

Cloud Sequence for Cyclone/Hurricane/Typhoon

The cloud sequence of cyclones/hurricanes/typhoons is similar to that of a middle-latitudes (30°N–55°N) warm front. Cirrus clouds converge towards the storm destination, followed by cirrostratus clouds. Below them are altostratus and then stratocumulus clouds. Once the barometric pressure begins to fall rapidly, a wall of heavy dark cumulonimbus clouds appears. This is called the 'bar of the storm'.

Below is the cloud sequence for a cyclone/hurricane/typhoon:

- cirrus
- cirrostratus
- altostratus
- stratocumulus
- cumulonimbus (Bowditch 2002).

Determining Cloud Level

A good way to determine the level of cumulus clouds is to assess the size of the individual cloud elements. Low-level cumulus clouds are about the same size as, or larger than, your fist held at arm's length. One exception to this rule is when a small cumulus cloud is developing or evaporating. In that case, its direction or speed of motion may indicate that it is in the same layer as nearby larger cumulus clouds. Mid-level cumulus clouds are further away and the individual cloud pieces appear substantially smaller, about the size of your thumb at arm's length. High-level cumulus clouds are smaller still, with individual cloud pieces about the size of the nail on your little finger at arm's length.

Stratus clouds have no distinct cloud pieces to measure. For these clouds, a general rule is that cloud opacity tends to decrease with height. Thus, low-level clouds are generally thicker than mid-level clouds, and a high-level cirrostratus is very thin. Thus, by observing how much the cloud obscures the sun, you can estimate the level of a stratus cloud.

If there is precipitation, the chances are very good that you are dealing with a low-level cloud. Mid-level clouds occasionally precipitate, but this is a rare occurrence (S'COOL/NASA 2016).

Calculate the Cloud Base Height

The approximate cloud base height in kms is obtained by subtracting the dew point temperature from the outdoor temperature and dividing by 8.

Cloud base (in kms) is $(t_a - t_d) / 8$
where t_a is surface air temperature
t_d is dew point temperature

For example:

given the surface air temperature is 24.0°C
and dew point temperature is 11.0°C
cloud base height = $(t_a - t_d) / 8$
= $(24 - 11) / 8$
= 1.63 kms

Cloud Sequences in a Nutshell

Cyclone	Thunderstorm
cirrus	cirrus
cirrostratus	cirrostratus
altostratus	cumulonimbus
stratocumulus	
cumulonimbus	

Cold Front	Warm Front
cirrus	cirrus
cirrocumulus	cirrostratus
altocumulus	altostratus
cumulus	stratus
cumulonimbus	nimbostratus
stratocumulus	

Table 8: *Cloud sequences in a nutshell.*

5

Why a Small Increase in CO$_2$ Critically Affects Climate

Carbon dioxide (CO$_2$) is a powerful trace gas. Its concentration in the atmosphere remained at its normal interglacial *maximum* level of 280 parts per million (ppm) for the many previous interglacial periods, from 1 million years ago until 1750 when the Industrial Age began. In the 1 million years before the Industrial Age, it dropped to its normal *minimum* of 180 ppm during the many ice ages.

In 2022, it reached 420 ppm. This is an increase of over 50% on the maximum, which had remained unchanged for 1 million years until 1750. CO$_2$ concentration of about 400 ppm will cause global warming of 2°C (3.6°F) and 450 ppm will cause global warming of 3°C.

So why does such an apparently small increase in CO$_2$ content in the atmosphere make a critical difference to the global surface temperature of the Earth, i.e. to Earth's climate?

The Earth is accumulating heat equivalent to four Hiroshima bombs of heat per second.

The atmosphere is transparent to visible radiation from the sun. When the sunlight is absorbed by the Earth's surface, it is re-emitted as heat (infrared radiation), and is almost entirely absorbed—over 90 %—by the CO_2 in the atmosphere. The CO_2 instantly re-emits the heat in all directions, heating the surrounding atmosphere. Most of the CO_2, around 60%, goes downwards towards the lower atmosphere and ground.

Although water vapour and clouds together absorb 75% of the Earth's heat radiation (Lacis 2010), they cannot determine the temperature of the atmosphere. Water vapour and clouds depend on temperature and air circulation in ways that CO_2 does not. They condense and cannot maintain a temperature structure for the atmosphere. CO_2 accounts for 80% of the non-condensing gases that maintain the temperature structure of the Earth, acting as the control knob of the Earth's thermostat. It controls the amount of water vapour and clouds.

The Earth is absorbing 0.5 Watts/m^2, more than it is radiating to space (Hansen 2012). As we add more CO_2 to the atmosphere, this absorption increases. If we multiply this rate by the surface area of the Earth (5.100656×10^{14} m^2), we find that the Earth is accumulating heat at a rate of 2.6×10^{14} Watts (or Joules per sec).

No Slow Down in Global Warming

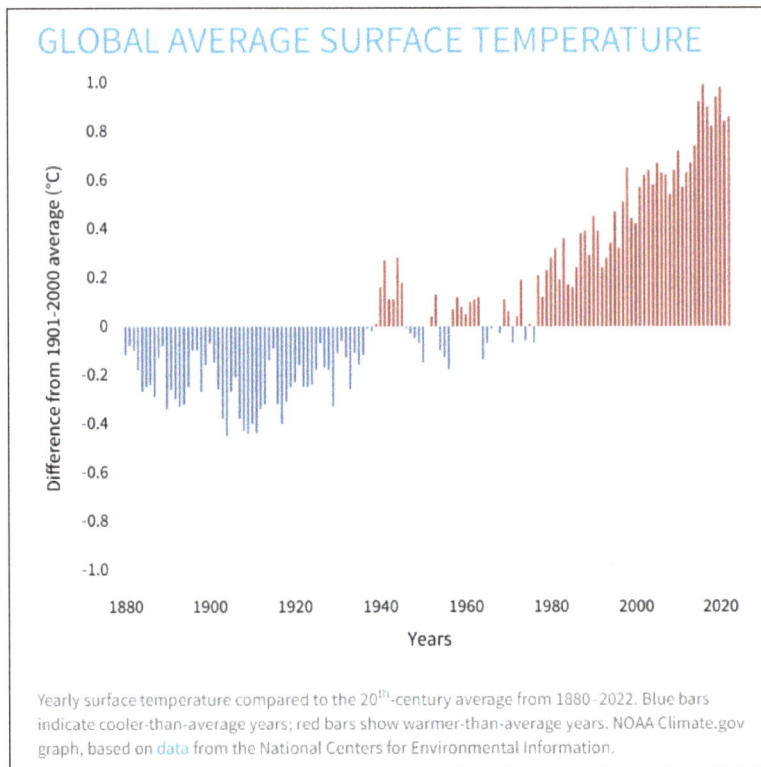

Figure 45: *Data showing no recent slowdown in global warming.*

Given the Hiroshima atomic bomb yielded an explosive energy of 6.3×10^{13} Joules (Malik 1985), this is equivalent to ***four Hiroshima bombs of heat per second***. Understandably, this compounds over a decade or more. Since 1998, our climate has absorbed over 2 billion such bombs in accumulated energy.

Our climate in 2013 absorbed the equivalent of 126,227,704 bombs in accumulated energy. Humans added a further 36 billion tons of CO_2 in 2013—the hottest year ever recorded until then in Australia—but the natural systems could only handle 15 billion tons of CO_2 per year, so the rate of heating increases every day. As the ocean absorbs over 90% of carbon dioxide, it becomes more acidic.

Combined with increasing ocean temperatures, this diminishes its ability to continue absorbing CO_2. Humans added 36.6 billion tons in 2022, according to the Global Carbon Project at NOAA.

On 9 January 2014, the Niagara Falls partially froze over. How did global warming make this extreme weather event even worse? The jet stream or polar vortex that forms a circular band of high-speed winds (160 km/h or 100 mph) around the Arctic region holds the colder Arctic air in place, protecting the region further south. As the Earth warms and the Arctic sea ice melts, the polar region warms faster than the region further south towards the Equator. This causes the jet stream to *slow down* and meander, pushing further south than usual and bringing freezing cold weather to North America.

Why is a Change in the Earth's Global Average Temperature a Big Deal?

The global *average* temperature of the Earth's atmosphere combines the temperatures of all places on Earth. Globally, a rise of 1°C makes a large difference. Even a small rise in Earth's global average temperature means melting ice at the Poles and a sea level rise. Some places flood while others have droughts. The extra heat is driving regional and seasonal temperature extremes, reducing sea ice alarmingly around the Poles, and intensifying heavy rainfall. The amount received in an ordinary rain shower in south eastern Australia has doubled in recent years. Habitat ranges for plants and animals are changing in response to climate change. And that's why a change in the Earth's global average temperature is a big deal.

It is important to remember that it is the *average* temperature of the Earth's atmosphere that determines its *behaviour*, *not* the day-to-day fluctuations of the temperature. Global warming driven by the burning of fossil fuels such as coal, oil, and gas is causing the *average* temperature of the atmosphere to keep rising.

Evidence for Rapid Climate Change is Compelling

According to NASA, the evidence for rapid climate change is compelling:

- Global average temperature is rising.
- The ocean is getting warmer.
- 90% of the extra energy in the ocean.
- The ice sheets are shrinking.
- The Greenland and Antarctic ice sheets have decreased in mass.
- Extreme events are increasing in frequency.

Glaciers are Retreating

Glaciers are retreating almost everywhere around the world—including in the Alps, Himalayas, Andes, Rockies, Alaska and Africa.

Snow Cover is Decreasing

Satellite observations reveal that the amount of spring snow cover in the Northern Hemisphere has decreased over the past five decades and the snow is melting earlier.

Sea Levels are Rising

The rate in the last two decades, however, is nearly double that of the last century and accelerating slightly every year.

Arctic Sea Ice is Declining

Both the extent and thickness of Arctic sea ice has declined rapidly over the last several decades.

Ocean Acidification is Increasing

Since the beginning of the Industrial Revolution, the acidity of surface ocean waters has increased by about 30%. This increase is due to humans emitting more carbon dioxide into the atmosphere and hence more being absorbed into the ocean.

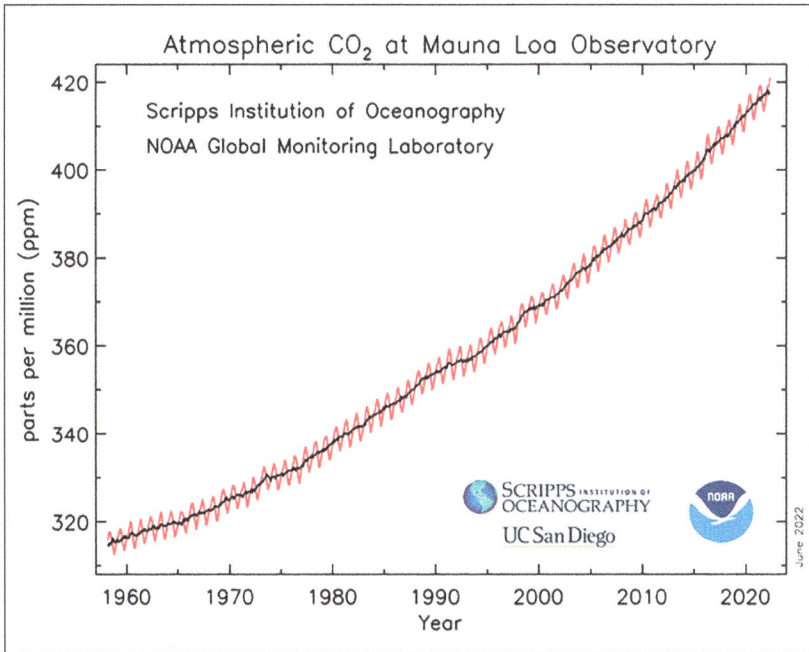

(NOAA, 2022) The diagram above was downloaded 3 June 2022 (https://www.noaa.gov/news-release/carbon-dioxide-now-more-than-50-higher-than-pre-industrial-levels).

Stronger Storms in a Warmer World

Over the last 35 years, the number of severe storms worldwide has doubled. Their wind speed and destructive potential has increased significantly.

> *Most of our weather comes from the oceans. Climate change caused by global warming of the sea surface layer means the largest storms will be fewer but more intense, with greater maximum wind speed and destructive potential.*

Storm updraft speeds are strengthened by about 1 m/sec with global warming, when a computer simulation is done using twice the present CO_2 concentration in the lightning-producing regions over land (Del Genio *ibid*).

'Moisture drives the engine, it is the fuel of these storms.'
(McGuire, 2019)

More moisture from global warming allows for more condensation and more latent heat release. The strongest storms can use all the extra latent heat released, creating a more vigorous updraft, and for the strongest storms the wind speed is increased further.

When CO_2 and other greenhouse gases trap heat in the atmosphere, about half of this heat is redirected back to the land and ocean surface. The top few metres of oceans store as much heat as the whole atmosphere. Most of our weather comes from the oceans. The global sea surface temperature as a result of climate change is more than 1°C higher now than 140 years ago. Most of this temperature increase has occurred since 1960. This warmer surface water dissipates more readily into water vapour, facilitating the growth of storms. The additional moist air provided by the warmer surface water, being lighter than drier air, rises faster, further increasing the rate of the central pressure fall of the developing storm. Storm growth requires an increase in the sustained wind speed and this is driven by the rate of pressure fall. Therefore, global warming of the sea surface layer increases the rate of pressure fall and consequently, the rate of sustained wind speed (storm intensity) increases.

In conclusion, climate change caused by the global warming of the sea surface layer means that the rate of pressure fall will be larger than it would be otherwise. **This means the threshold for storms will be reached sooner.** The storms will be more intense, with greater maximum wind speed and destructive potential.

Global warming has meant that the Earth's atmosphere needs to process 7% more water vapour from every 1°C rise in its average temperature. That is, warmer air has a larger capacity to hold moisture according to the Clausius-Clapeyron equation from physics. This forces more of the atmosphere's total energy into the processes of rain, snow and evaporation, but there is less energy for atmospheric circulation, resulting in fewer storms. The extra water vapour and associated latent heat will increase the intensity of the remaining storms, especially the largest ones.

> *Global warming has meant that the Earth's atmosphere needs to process 7% more water vapour from every 1°C rise in its average temperature.*

There is a super-charging of rainfall rates with temperature for tropical cyclones because global warming is increasing the tropical cyclone intensity itself. Consequently, we get an increase in water vapour from the environment (Clausius-Clapeyron equation) and even more rain from Global Warming induced tropical cyclone intensity.

The last three decades have seen an increase of about 15% in tropical storm intensity. This represents an increase in the destructive potential of these storms of 52%. The number of severe storms worldwide has doubled in the last 35 years.

The decade to 2017 saw a five-fold increase in the number of tornadic supercell thunderstorms in Australia compared to the previous decade. Oceans absorb 93% of the excess heat from global warming. This is increasing the length of underwater heatwaves.

World Population Growth is the 'Elephant in the Room'

Records show that CO_2 emissions from the burning of fossil fuels respond proportionately to increases in population size. Human population grew from 1.6 billion to 6.1 billion in the last century. CO_2 emissions grew twelve-fold fold during the same time. World population growth is the 'elephant in the room' when it comes to global CO_2 emissions and hence global warming.

How was Global Warming Discovered?

The following answers the challenging question posed on page 4 which is:

The planet Venus has the highest surface temperature in the Solar System yet it is not the closest planet to the Sun. Why is it so?

According to NASA (http://climate.nasa.gov/evidence), in the 1860s, physicist John Tyndall recognized Earth's natural greenhouse effect and suggested that slight changes in the atmospheric composition could bring about climatic variations. But it was not until 1988 that a long-term warning given by Dr James Hansen raised awareness significantly. He was responsible for an experiment aboard the Pioneer Venus Mission of NASA when he realised that the most important change was the burning of fossil fuels on Earth. He already knew that carbon dioxide determined the climate on Venus. The planet Venus has the highest surface temperature in the Soler System yet it is not the closest planet to the sun. The atmosphere of Venus contains 97% carbon dioxide gas and is hot enough to melt lead. He realised that the rapidly increasing carbon dioxide content in the Earth's atmosphere would surely affect the Earth's climate.

On 23rd June 1988, Dr Hansen said to the US Senate Committee that he had '99% confidence … Earth was being affected by human-made greenhouse gases, and the planet had entered a period of long-term warming.'

6

Windstorms

A windstorm is a storm with winds exceeding 55 km/h (34 mph). Wind can come in gusts or as sustained strong winds. Typically, the wind is not accompanied by precipitation (rain). There are hot windstorms. With fire, they are called firestorms; with sand, sandstorms; with dust, dust storms. Dust storms occur in arid and semi-arid regions where loose dust is blown from the surface. The Australian Outback red dust storms are huge. They can be 500 km (310 mi.) wide. Sandstorms occur when a whole layer of sand is lifted at a time.

Figure 46: *Dust storm approaching Stratford, Texas, U.S.*

7

Snowstorms

"Climate change is fuelling an increase in the intensity and snowfall of winter storms. The atmosphere now holds more moisture, and that in turns drives heavier than normal precipitation, including heavier snowfall in the appropriate conditions" (Trenberth 2011).

A front is "a boundary or transition zone between two air masses of different density, and thus (usually) of different temperature. A moving, e.g., cold front if colder air is advancing" (US Department of Commerce National Oceanic and Atmospheric Administration 2009).

Winter storms can form with the arrival of a cold front when a mass of cold, dry air moves into a mass of warm, moist air. Gusty winds and a rapid temperature drop often result. Snow, sleet or freezing rain can occur when the ground temperature is low enough.

Blizzards are severe snowstorms with a sustained wind speed of 56 km/h (35 mph) or more, lasting for 3 hours or more, giving a large snowfall and visibility of less than 0.4 km (0.25 mi.). Where the winds associated with an extratropical cyclone are strong, a blizzard can occur.

The heaviest snowfall sometimes occurs where an extratropical cyclone's air is forced to rise over the mountains, causing uplift of the air to a higher altitude. Snowstorms can occur in very different conditions when very cold air flows over a large ice-free lake and collects water vapour. Sometimes, a cold air front from the Arctic can travel south. The cold front can carry humidity for a long distance in winter months, producing snowstorms.

As with any storm, **snowstorms are all about a fall in barometric pressure**. In the Northern Hemisphere, a blizzard can form when the approach of a low-pressure system occurs with below-zero temperatures. **You can predict snowstorms, including blizzards, using a barometer** and observing that the barometric pressure is below its ceiling value and the rate of change exceeds a threshold.

In February 2013, Winter Storm Nemo dropped 29 mb within 24 hours, which made it a weather bomb.

Figure 47 (below): *The 2016 snowstorm in Washington D.C. ranked as a category 4 storm on the NESIS scale.*

Conclusion

General weather is affected by many factors, including pressure, temperature, wind speed, and humidity. Fortunately, a storm's maximum wind speed (intensity) depends *entirely* on its central pressure, as shown in Appendix 2 on page 107. That is why the barometer is such a powerful instrument for predicting storms.

Your barometer lets you be more aware of and connected to nature. The number of rainy days each year vastly exceeds the number of days with a storm, so this book shows you how to predict rain as well as storms. Predicting storms is easy and you will soon gain confidence in doing so. All storms, whether they are thunderstorms, dust storms, firestorms or snowstorms, are governed entirely by the barometric pressure and its rate of change. It is best to be your own meteorologist and not rely on forecasts coming from other than your immediate vicinity. All you need is your barometer or barograph app that uses a pressure sensor and the reliable barometer rules given here. If a simple approach works, it is usually better to adopt it.

Of course, it is useful to keep an eye on the sky to confirm your predictions. Cloud arrangements reflect surface wind direction and the distribution of atmospheric moisture. Cloud sequences give a reliable 24-hour forecast. Learn to identify cirrus clouds, although cirrus clouds alone do not mean change. It is the sequence of different cloud types that follow them that can indicate a change. If they appear to be moving, they must be travelling very fast and any change will come soon. If they start to thicken and lower, then check your barometer. Similar cloud patterns produce the same weather each time they occur. It is worthwhile becoming familiar with 10 cloud types and then focus on cloud sequences. Study the photographs of clouds given at http://www.srh.noaa.gov/jetstream/clouds/cloud-chart.html#myModalh5

You always need to stick with the basics and ask the question: Is the current pressure less than 1009 mb? If not, it is *not* a storm. If pressure is less than 1009 mb, has the pressure fallen 4 mb or more in the last 3 hours? If so, an ordinary thunderstorm is approaching. If there is no storm onset it means a *Severe Thunderstorm* is developing. The rate of pressure fall of a *Severe Thunderstorm* typically will exceed *The Thunderstorm Rule threshold* from the outset with no sign of storm onset for 8 to 12 hours.

What is the size and intensity of the coming storm? Thunderstorms quick basics are given in Figure 34, page 59 and the rules for pre-dicting all storms are given in Figure 35, page 60 for easy reference.

Being able to use your barometer can be a real game-changer. Up to as many as 500,000 people worldwide can die in a large storm in a single year. *Your barometer will give you at least 24 hours ear-ly warning of an approaching hurricane making landfall.* This is because the pressure becomes almost steady for 6 hours at least 24 hours before it makes landfall.

Barometer readings need to be taken regularly, usually every 6 hours but even more frequently at sea. A wind speed greater than 15 knots (28 km/h) can capsize a small boat. We have shown that a minimum 6 mb fall in barometric pressure over 3 hours will produce a strong wind of 29.3 knots (54.2 km/h).

You may wonder, when will you be safe from storms?

Safe conditions are:

1. If the pressure remains above 1009 mb, there can be no storms.

2. If the pressure remains above 1005 mb, there can be no severe thunderstorms.

3. If the pressure remains above 1000 mb, there can be no cyclones, hurricanes or typhoons.

4. If the pressure falls less than 3 mb in the last 3 hours, there can be no storms.

When the pressure starts falling at more than 1.0 mb / hour a storm may be approaching. Apply the Thunderstorm Rule (i.e. when barometric pressure falls 4 mb or more in the last 3 hours to below 1009 mb, a storm is approaching) to be certain.

You can have at least 5 hours early warning of a deadly tornado from its rigid straight-line signature on a barograph.

You can have 24 hours early warning of a Hurricane in the tropics when the pressure has been almost steady for 6 hours.

There will be times, however, that allow less frequent monitoring of barometric pressure. For example, the barometer is steady when wind speed is steady. If the wind speed is steady, the current weather you are observing is likely to persist, and less frequent barometer readings will be needed.

You can also use an altimeter for predicting storms or rain.

Currently, 13 minutes lead time is all that can be given before a tornado strikes. The Tornado Early Warning Rule presented for the first time in this book is ground-breaking and will save lives. The early warning using the jump in lightning flash rate with a developing tornado is from the cutting edge of research. *You can now have at least 5 hours early warning of a tornadic supercell thunderstorm. A barometer will detect a supercell thunderstorm long before it is visible to radar or satellite.*

The Early Warning rule for bush firestorm conditions (see page 55) needs to be immediately adopted by firefighters and weather forecasters.

The author suggests that 1-hour early warning of a tornado from a non-supercell is possible.

Many popular smartphones have an altimeter. If you **remain at the same altitude**, the altimeter will show an apparent increase in altitude with falling barometric pressure. I am familiar with the altimeter, having taught RAAF aircrew. Actually, aircrew cannot use an altimeter in this way as the aircraft would be frequently changing altitude, but anyone else can.

We have seen that an apparently small increase in CO_2 content in the atmosphere is making a critical difference to the global surface temperature of the Earth and its climate. We now know that the Earth is accumulating heat equivalent to four Hiroshima bombs of heat per second. The last three decades has seen an increase of about 15% in the intensity of cyclones/hurricanes/typhoons. The number of severe storms worldwide has doubled in recent decades. Total CO_2 emissions from fossil fuels continue to increase. Global temperatures could reach an irreversible tipping point by 2030 if the world does not take effective action to reduce CO_2 emissions now. **Our planet faces a Climate Emergency**.

We have learnt that the swell and the cloud sequence from a distant storm can help us predict a storm that is on its way. We now know that observing the cloud sequence and ocean swell from a distant storm is a valuable forecasting aid.

All you need to predict storms is your barometer or barograph app that uses a pressure sensor and the reliable barometer rules given in this book. You don't need a traditional forecaster, a satellite or a supercomputer.

Let this book and the companion app make your forecasting easy and fun.

APPENDICES

Appendix 1

Weather Lore Sayings

"Red sky in the morning, sailors take warning".

This old favourite can be traced back to a message in the New Testament. Jesus told them, "You have a saying that goes, ... *red sky at morning, sailors take warning.*" 'But its meteorological underpinning is that the red sky is caused by the dawn light bouncing off ice crystals in high cirrus clouds' (McGuire ibid, 2019).

The following weather lore sayings selected from the US National Park Service: https://www.nps.gov/grte/learn/education/class-rooms/upload/Weather-Lore-Sayings.pdf

"Mares' tails and mackerel scales Make lofty ships carry low sails."

Mares' tails are cirrus clouds, called this because they sometimes resemble the flowing tail of a horse in the wind. Mackerel scales are altocumulus clouds. They appear broken and scaly. Neither of these cloud types will bring rain or snow themselves. They do, however, precede an approaching storm front by a day or two.

"If clouds move against the wind, rain will follow."

Clouds that are moving in a direction that differs from the way the wind is blowing indicates a condition known as wind shear. This sometimes indicates the arrival of a cold front. Weather fronts usually bring rain.

"When the dew is on the grass, Rain will never come to pass."

When grass is dry at morning light, Look for rain before the night." Again, if there is no dew on the grass, it means the sky is cloudy or the breeze is strong, both of which may mean rain.

"If a circle forms 'round the moon, Twill rain soon."

The circle that forms around the sun or moon is called a halo. Halos are formed by the light from the sun or moon refracting (bending) as they pass through the ice crystals that form high-level cirrus and cirrostratus clouds. These clouds do not produce rain or snow, but they often precede an advancing low pressure system which may bring bad weather.

"If birds fly low, Expect rain and a blow."

When the air pressure is high, it is easier for birds to fly at a higher altitude. If the air pressure is low, indicating bad weather, birds can't fly as high because the air is less dense.

Appendix 2

Derivation of the Law of Storms

The maximum sustained wind speed is a measure of the maximum intensity of a storm. However, the central pressure values are more reliably measured than the sustained wind speed. The most widely used formula by Atkinson and Holliday (1977), relating the maximum wind speed (km/h) and the surface central pressure, p_c (hPa or mb), is given by:

Equation 1

$$v_m = 12.24(1010 - p_c)^{0.644} \quad (1)$$

Only a few real storms ever reach their maximum sustainable wind speed. For this and other reasons, a 20% adjustment in wind speed has been applied to Equation 1 giving Equation 2:

Equation 2

$$v = 9.8(1010 - p_c)^{0.644} \quad (2)$$

Transforming Equation 2 and including the adjustment we get the Law of Storms expressed in calculus:

Equation 3

$$\frac{dv}{dt} = -\frac{22.3}{v^{0.553}} \; X \; \frac{dp}{dt} \qquad (3)$$

where v is the wind speed in km/h, and p is the central pressure in hPa or mb.

Storm growth requires an increase in the sustained wind speed given by the acceleration term $\frac{dv}{dt}$ and this is driven by $-\frac{dp}{dt}$ *the rate of pressure fall* in Equation 3.

The Law of Storms states:

> *At a given wind speed, the marginal rate of increase of sustained wind speed is directly proportional to the marginal rate of pressure fall.*

Equation 2 above implies that the term 'at a given wind speed' can be replaced by 'for a given pressure' and the Law of Storms can be restated:

> *The marginal rate of storm growth is directly proportional to the marginal rate of pressure fall for a given pressure.*

Therefore, the criteria for a storm can be specified by:

- a pressure ceiling value, and
- a threshold rate of pressure fall.

For example:

Ordinary Thunderstorm Warning
Last 3 hours ▼ 4 mb or more (current pressure < 1009 mb)

This means if barometric pressure has fallen 4 mb (or more) to below 1009 mb in the last 3 hours, a storm is approaching.

Appendix 3

Types of Barometers

Torricellian barometer

The original Torricellian barometer consisted of an inverted glass tube standing in a bath of liquid. In 1643, Torricelli constructed a tall water barometer that protruded through his roof. Such a barometer needed to be at least 10.3 m (33.9 ft) high.

Mercury barometer

The mercury barometer is less than a tenth of the size of a Torricellian barometer. It works by balancing the weight of mercury in an inverted tube against the pressure of the atmosphere pushing down. The height of the mercury in the vertical tube is a measure of the atmospheric pressure. The weight of a column of air pushing down on a 1-m-square area of surface at sea level is about 72 kg.

In the diagram below, the pressure at X supporting the column of mercury is equal to the force per unit area exerted against the surface of mercury by the atmosphere at Y.

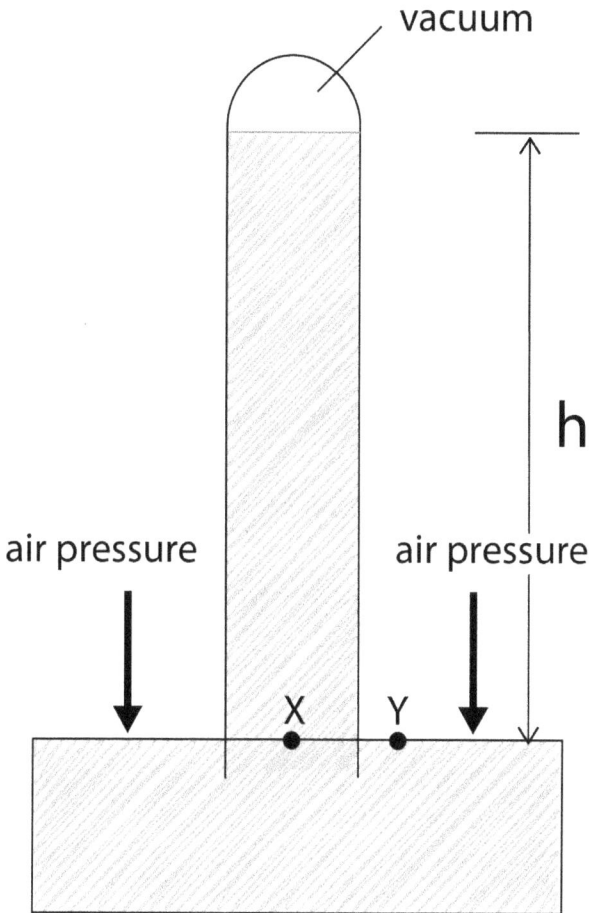

Figure 48: *Mercury barometer.*

Pressure at X = ρgh where ρ is the density of mercury, g is the acceleration due to gravity and h is the height of the column of mercury.

Aneroid barometer

The aneroid barometer consists of a flexible box where the air inside the box has been removed. The changes in the pressure on the box by changing air pressure causes it to expand or contract and move the needle on a gauge.

Digital barometer

Digital barometers work in a similar manner to an aneroid barometer. An electronic pressure sensor converts pressure on a sealed cavity made of semiconductor material into an analogue electrical signal. An analogue-to-digital converter gives digital values of pressure.

Conversion of pressure units

1.0 Atm = 1013.25 hPa = 1013.25 mb = 29.92 inHg

Pressure sensor on smart phones - a breakthrough

The inclusion of a pressure sensor on good smart phones was a breakthrough in predicting storms and a real game changer. Apps such as *Marine Barograph* can now read accurate pressure values at your immediate location. Often forecasts from the various agencies are based pressure readings taken too far away from the storm's pressure centre and are consequently unable to reliably predict storms. The simple rules given in this book when applied to the accurate pressure values taken by your phone's pressure sensor at your immediate location are reliable. Anyone now can predict storms using the pressure sensor and a barograph app on their smart phone.

Careful use of your device

Always keep your device *still* when monitoring the pressure. The barometer or Smartphone (with its pressure sensor) needs to be kept away from the extremes of humidity and temperature.

Appendix 4

Storm Records

Largest	
Super Typhoon Tip	2,220 km (1,380 mi) diameter, NW Pacific (Landsea 2010)
Most intense	
Super Typhoon Tip	870 mb (25.69 in Hg), NW Pacific (Landsea 2010)
Storm surge	
Tropical Cyclone Mahina	14.6 m (48 ft) 5 March 1899 Bathurst Bay Australia (Masters 2012)
Highest wind gust	
Cyclone Olivia	408 km/h (253 mph) 10 April 1996 Barrow Island (Courtney et al. 2012)
Largest eye	
Typhoon Carmen	370 km (230 mi.) 20 August 1960 Northwest Pacific (Lander 1999)

Longest tornado track	
Tri-state tornado	352 km (219 mi.) 18 March 1925 Missouri (US Department of Commerce National Oceanic and Atmospheric Administration 2009)
Longest lightning strike	
Storm	Lasted 7.74 seconds 2012 Alpes-Côte d'Azur Provence, France (Gray 2016)
Longest distance travelled, lightning	
Storm	Horizontal distance 321 km (200 mi.) June 2007 Oklahoma, US (Gray 2016)
Worst firestorm	
Black Saturday bushfire	Triggered before north wind arrived. Pressure drop of 10 mb sustained for at least 6 hours. Temperature > 40°C for 6 hours, peaked at 46.4°C. 7 February 2009 Kinglake, Australia
Worst blizzard	
Blizzard of 1972	Blizzard dropped up to 7.9 m (26 ft) of snow from 3–8 February 1972 Iran (NY Times 1972)
Worst windstorm	
Pacific Northwest windstorm	Columbus Day Storm, Wind gusts > 233 km/h (145 mph). Some reports give peak wind 288 km/h (179 mph). 12 October 1962 Cape Blanco, Oregon (LaLande 2017)
Largest dust storm	
Eastern Australian Dust storm	500 km (310 mi.) wide, 1,000 km (620 mi.) long. From 22 to 24 September 2009, Australia (Malkin 2009)

Appendix 5

Doubling the CO$_2$ Content

Let's calculate the temperature for doubling the atmospheric CO$_2$ content using the *Equation for Global Warming*:

$$\Delta T = 1.66 \, ln \, (C/C_0)$$

The equation above by the International Panel on Climate Change (IPCC) gives the temperature increase above the average global surface temperature, ΔT (in °C) when CO$_2$ concentration increases from C_0 to C ppm.

Let us calculate the temperature increase for instant doubling the atmospheric CO$_2$ content when there is no feedback.

We calculate ΔT by setting $C = 2 \, C_0$ in Temperature Increase Equation: $\Delta T = 1.66 \ln (C/C_0) = 1.66 \times 0.693 = 1.2$°C

The increased surface temperature from the instant doubling of CO$_2$ content allows an increased water vapour content by maintaining a constant relative humidity. The extra water vapour increases the overall absorption by water vapour itself raising the surface temperature further by about another 1.2°C. The total increase is about 3°C when all feedbacks are included.

Appendix 6

The Thunderstorm Rule in a Nutshell

One simple rule, *The Thunderstorm Rule*, gives you early warning of any storm that is approaching your immediate location.

When barometric pressure falls 4 mb or more in the last 3 hours to below 1009 mb (The Thunderstorm Rule) an **ordinary thunderstorm** is preparing to onset in the usual way, the pressure continues to fall to a minimum and then jumps, the storm then finally onsets within 2 hours.

When the threshold for an ordinary thunderstorm is passed without storm onset and there is an ongoing steady pressure trend, a Severe Thunderstorm is developing. Every hour that the pressure falls without storm onset the storm is intensifying.

The rate of pressure fall of a *Severe Thunderstorm* typically will exceed *The Thunderstorm Rule* threshold from the outset with no sign of storm onset for 8 to 12 hours.

Island in the Storms:
A short story

By Robert Ellis

Figure 49 (below): *Storm cloud near Bundeena, Australia.*

I am the only resident, except for my Golden Retriever, on a small island off the northern coast of Queensland in Australia. The island, which is about a metre above the high- tide mark, has access to the mainland over a bridge. The bridge is about a half-hour drive from my cottage.

We're subjected to plenty of unsettled weather here and so we're no stranger to tropical cyclones, storms, rough seas and high winds. Hence the reason I named my dog Squall, especially because she was born during a particularly Severe Thunderstorm. Her poor mum only lasted six months after Squall was born, when she developed septicaemia from a bad cut on the rocks down at the beach.

I've been living here for several years now and so I'm used to the weather patterns here. But when I first moved here, it was a different story and I almost didn't live to tell it. I've always kept a journal and I recorded the events of my first Severe Thunderstorm on the island. Since that time, I've become a bit of a weather freak—probably that's quite normal when you live in a place subjected to the elements—and so I'm able to tell you the story using all the correct weather terminology.

One thing I have had right from the beginning since I began living here is a weather app on my smart phone. If I'd understood then what I know now, I'd have taken proper notice of it, instead of putting myself or Squall at risk.

But if I'd done so, there wouldn't be a story to tell today. These are my adapted journal entries.

* * *

I woke on Saturday morning, hoping for a better day than yesterday. Yesterday the sky at sunrise was a deep fiery red. It was covered with low clouds and the driving rain was persistent. A fresh gale was blowing in gusts. This was not unusual because it was always windy on the island. The waves near the island were about six metres high with streaks prominent.

I checked the electric wall clock: 5.45 am. The amount of light told me it was much later than that. My wristwatch said 11.45 am. 'Mmm,' I thought, 'no power.' Unless there was a miracle, I'd have no phone once my battery went flat, which wouldn't be much longer because I'd forgotten to charge it during the night. I'd also have no Wi-Fi, no internet and no radio. Nil communications until power was restored, whenever that might be. This was frustrating, as the power had been out yesterday as well and was only restored 12 hours ago.

When I switched on my smart phone, I noticed to my alarm that the display read: Cyclone/Hurricane/Typhoon Wind Warning. I had no idea how long the app had been sending me the warning. Because I'd switched off the phone, all the previous notifications had been wiped.

I inserted my hearing aids, as I am practically deaf without them, and immediately heard the howl of the hurricane outside and the rain hammering against the windowpanes. Donning my wet-weather gear, I went outside. The wind was blowing the rain almost horizontally and I had to squint in order to see anything through it. I also had to grasp onto the railing to be able to stand upright. A savvy friend who had also lived in remote locations subjected to wild weather had suggested I install it so that I could as least continue with some limited outdoor functions in severe weather. I was grateful for it now. Squall had come outside as well and she was staying close to me, whimpering against my legs. Dogs have a sixth sense. I wish I'd asked Squall what was coming.

'Look at that sea,' I said to her. The ocean surface was completely covered with white foam. I clicked on the windmill icon button on my app, which displayed: Air filled with foam and spray, with low visibility, indicating at least hurricane-force winds.

The storm surge had advanced to the very edge of both sides of the road that led off the island, and furthermore, it was advancing with every wave. I knew that the storm surge could be up to six metres.

What I didn't know was how much time I had before the hurricane approached landfall and the storm surge covered the coastal road.

I'd heard about this happening and although I hadn't experienced it myself, I guessed I might still have about four hours.

Nevertheless, I wasn't about to take any chances. I'd already been too laidback about this weather event.

I ran around the cottage, securing the exterior storm shutters over the windows. Then I sprinted inside, turned down the flue on the wood burner, grabbed my car keys off the hook near the kitchen and slammed the cottage door shut, hoping my little home would withstand the high winds and driving rain. Opening the driver side car door, I urged Squall in, although she needed very little persuasion. 'Jump in, girl!' I shouted above the wind, pushing her into the front passenger seat. She barked and immediately obeyed, while I jumped into the driver's seat and put the key in the ignition faster than I'd ever done before. Flooring the accelerator, I sped away, increasingly desperate to leave the island while we still had time.

The tyres sprayed up water on both sides of the car and I was concerned about hydroplaning, but I couldn't let up on the acceleration, even if it meant I lost a little control. I'm a good driver and we had to get out of there fast.

The windscreen wipers were on rapid but even so, I could barely see through the windscreen in the driving squalls. It felt as if the howling wind was going to lift us into the sky at any moment, like Dorothy in the Wizard of Oz twister, and I was finding it difficult to keep the car straight on the road.

After what seemed like a lifetime, negotiating branches that had fallen across the roadway and frequently and anxiously peering in the rear-vision mirror at the road behind us to make sure the water wasn't approaching us unawares, I finally spotted the bridge through the driving rain. Cautiously approaching, and checking to see that the waters beneath the bridge had not risen to dangerous levels, I drove across it onto higher ground on the opposite side.

There was no other traffic about and I stopped and parked by the roadside for a moment, watching the rain becoming heavier and the wind increasing in strength. I could see the low-lying areas on the island gradually flooding as I watched. We had escaped just in time. The wind was screaming, as if enraged that we had made our get-away safely. In the distance, the sea was white with foam and streaks. Huge waves were crashing onto the beach and not for the first time, I was relieved I'd made the decision to build my cottage a couple of kilometres back from the shore and up on higher ground.

I knew what was coming next—the eye of the storm—and while I can describe it, I sure wasn't going to wait around for it.

I knew that the worst thing I could do was to get out of the car, or even stay where I was much longer, while the calm eye passed over, as I would soon be caught off guard by the violent winds in the opposite eyewall.

Turning to Squall, who hadn't stopped whimpering since before we'd left the cottage, and who was now looking at me with eyes that pleaded with me to get her to safety, I stroked her head and said, 'Too right, girl. Let's get out of here, now!'

* * *

Figure 50 (on following page): *Lifeguard chair on beach.*

Months later, after the close call of nearly being trapped by the cyclone, I was looking forward to a spot of fishing off the island in my little 12-footer.

The weather forecast on the radio indicated that fine weather was ahead. Fine weather is essential for fishing in such a small boat. Waves from a Severe Thunderstorm can easily swamp the boat, as the storm's sustainable wind speed is at least 56 km/h. That's a near Gale-Force wind.

I loaded the gear and headed out across the smooth sea. The fishing grounds were about two hours away, which feels a long way by boat. It was good to arrive.

I was just starting to get a few bites when I noticed that the barograph on my storm buddy app had fallen 1.57 mb in the last hour indicating that a storm may be approaching. Pressure fell 4.7 mb in 3 hours exceeding the *The Thunderstorm Rule* threshold but there was no sign of storm onset and the pressure continued to display an ongoing steady pressure trend. A Severe Thunderstorm was developing. According to the *Severe Thunderstorm Early Warning Rule* I had at least 5 hours (and possibly up to 9 hours) early warning of the onset of a Severe Thunderstorm. I had plenty of time to get back to shore and it took about 2 hours.

I have to thank my smart phone and that little barograph app, **Marine Barograph** which can be used on land and sea.

Always remember that if The Thunderstorm Rule (page 116) is applied and there is no storm onset it means a Severe Thunderstorm is developing.

References

Atkinson, GD & Holliday, CR 1977, 'Tropical Cyclone Minimum Sea Level Pressure/Maximum Sustained Wind Relationship for the Western North Pacific', *Monthly Weather Review*, vol. 105, pp. 421-27.

Atmospheric CO_2 at Mauna Loa Observatory, NOAA Global Monitoring laboratory. Uploaded 3 June 2022 https://www.noaa.gov/news-release/carbon-dioxide-now-more-than-50-higher-than-pre-industrial-levels

Borg, CY 2015, 'The Importance of Period. Swell period: key wave predictor', viewed 26 June 2017, http://www.surfline.com/surf-news/the-importance-of-wave-period_125149/

Bowditch, N 2002, *The American Practical Navigator*, Paradise Cay Publications, United States Government National Imagery and Mapping Agency, Washington.

Bunkers, MJ, Hjelmfelt, MR, Smith, PL 2006, 'An observational examination of long-lived supercells: Part I characteristics, evolution and demise, *Weather and Forecasting*, vol. 21, p. 682, Table 2.

Courtney, J, Buchan, S, Cerveny, RS, Bessemoulin, P, Peterson, TC, Rubiera Torres, JM, Beven, J, King, J, Trewin, B, Rancourt K 2012, 'Documentation and verification of the world extreme wind gust record: 113.3 m s–1 on Barrow Island, Australia, during passage of tropical cyclone Olivia', *Australian Meteorological and Oceanographic Journal*, vol. 62, viewed 20 July 2017, http://www.bom.gov.au/jshess/papers.php?year=2012

CSIRO Information Technology Group, 1997, *CSIRO Weatherwall (Aspendale Victoria)*, viewed 7 February 2009, http://webas-stage.dar.csiro.au/weatherwallDB/

Del Genio, AD, Yao, MS, & Jonas, J 2007, 'Will moist convection be stronger in a warmer climate?', *Geophysical Research Letters*, vol. 34, L16703, doi:10.1029/2007GL030525, p. 2.

Dorst, N 2004, 'What are the Early Warning Signs of an Approaching Tropical Cyclone?', viewed 20 July 2017, http://www.aoml.noaa.gov/hrd/tcfaq/H5.html

Edwards, R 2010, 'Online Tornado Frequently Asked Questions', viewed 30 August 2017, http://www.spc.noaa.gov/faq/tornado

Gray, R 2016. 'Now that's a BIG shock! World's Longest Lightning Bolt Stretched for nearly 200 mi. across Oklahoma', viewed 22 June 2017, http://www.dailymail.co.uk/sciencetech/article-3791693/Lightning-bolts-Oklahoma-France-deemed-worlds-longest.html

Hansen, J, *Storms of My Grandchildren*, Bloomsbury Publishing, New York, 2009 Preface xv.

Hansen, J, Sato, M. Kharecha, P & von Schuckmann, K 2012, 'Earth's Energy Imbalance', viewed 27 July 2017, https://www.giss.nasa.gov/research/briefs/hansen_16/

Jensen, P 2010, 'Eagle–Old World Wisconsin Tornado', viewed 07 September 2017, https://www.weather.gov/mkx/062110-eagle-oldworld-wi-tor

Lacis, AA, Schmidt, GA, Rind, D & Ruedy, RA 2010, 'Atmospheric CO2: Principal control knob governing earth's temperature', *Science*, vol. 330, issue 6002, pp. 356-59, DOI: 10.1126/science.1190653

LaLande, J 2017. 'Columbus Day Storm 1962', viewed 22 June 2017, https://oregonencyclopedia.org/articles/columbus_day_storm_1962

Lander MA 1999, 'A Tropical Cyclone with a Very Large Eye', *American Meteorological Society Pictures of the Month*, January 1999, viewed 2 August 2017, http://journals.ametsoc.org/doi/pdf/10.1175/1520-0493(1999)127%3C0137%3AATC-WAV%3E2.0.CO%3B2

Landsea, C 2010. 'FAQ: Hurricanes, Typhoons, and Tropical Cyclones', viewed 19 July 2017, http://www.aoml.noaa.gov/hrd/tcfaq/E1.html

Lydolph, PE 1985, *The climate of the Earth*, Rowman & Littlefield Publishers, Inc, Maryland.

McIlveen, R 1991, *Fundamentals of Weather and Climate*, Psychology Press, Oxford, UK.

Maddox, R. A. et al, 2013, Meteorological analyses of the Tri-State tornado event of March 1925, *Electronic J. Severe Storms Meteor*, 8(1), 1 – 27

Malik, JS 1985, 'The yields of the Hiroshima and Nagasaki nuclear explosions', *Los Alamos National Laboratory report number LA-8819*, viewed 27 July 2017, http://atomicarchive.com/Docs/pdfs/00313791.pdf

Malkin, B 2009, 'Largest dust storms in 70 years cover Sydney', 23 September 2009, viewed 22 June 2017, http://www.telegraph.co.uk/news/worldnews/australiaandthepacific/australia/6222210/Largest-dust-storms-in-70-years-cover-Sydney.html

Masters, J 2012, 'World Storm Surge Records', viewed 22 June 2017, https://www.wunderground.com/hurricane/surge_world_records.asp

McGuire, M, 'Storm Chaser', *SA Weekend, Adelaide Advertiser,* March 30-31, 2019, pp. 16-17. (ADV01Z01WE – VI)

National Aeronautics and Space Administration, Florida, viewed 6 October 2023, http://climate.nasa.gov/evidence

New York Times, 1972, 'Missing Put at 6,000 in Iranian Blizzard', 11 February, viewed 22 June 2017, http://www.nytimes.com/1972/02/11/archives/missing-put-at-6000-in-iranian-blizzard.html

Rogerson, T 2016, 'Determining Cloud Level, Cloud Tutorial Transcript', National Aeronautics and Space Administration, Florida, viewed 28 August 2017, https://scool.larc.nasa.gov/tutorial/clouds/cloudtypes_transcript.html

Sanders, F & Gyakum, JR 1980, 'Synoptic-Dynamic Climatology of the "Bomb"', *Monthly Weather Review*, vol. 108 (10), pp.1589–1606. Bibcode: 1980MWRv.108.1589S. doi:10.1175/1520-0493(1980)1081589:SDCOT>2.0.CO;2. 1 October 1980.

Schultz, C J, Petersen, WA & Carey, LD 2009, 'Preliminary development and evaluation of lightning jump algorithms for the real-time detection of severe weather. *Journal of Applied Meteorology and Climatology*, vol. 48, issue 12, pp. 2543–63.

Stano, GT, Schultz, CJ, Carey, LD, MacGorman, DR & Calhoun, KM 2014, 'Total lightning observations and tools for the 20 May 2013 Moore, Oklahoma, tornadic supercell', *Journal of Operational Meteorology*, vol. 2, issue 7, pp.71–88.

Trenberth, KE 2011, 'Changes in precipitation with climate change', *Climate Research*, vol. 47, pp 123-38. DOI:10.3354/cr00953.

US Department of Commerce National Oceanic and Atmospheric Administration 2009, 'Glossary', Page author NWS Internet Services Team, viewed 26 July 2017, http://w1.weather.gov/glossary/index.php?letter=f

US Department of Commerce National Oceanic and Atmospheric Administration 2009, 'Glossary', Page author NWS Internet Services Team, viewed 24 July 2017, http://forecast.weather.gov/glossary.php?word=squall

US Department of Commerce National Oceanic and Atmospheric Administration 2009, 'Longest Single Tornado Track 219 mi. (352 km) (Tri-state tornado, Missouri, IL, IN, F5 strength, March 18, 1925)', viewed 20 July 2017, http://www.crh.noaa.gov/Image/dvn/downloads/quickfacts_Tornadoes.pdf

US Department of Commerce National Oceanic and Atmospheric Administration (NOAA), National Data Buoy Center 2017, 'Measurement Descriptions and Units', viewed 8 August 2017, http://www.ndbc.noaa.gov/measdes.shtml

US Department of Commerce National Oceanic and Atmospheric Administration 2017, 'Thunderstorm Hazards–Tornadoes', viewed 8 August 2017, http://www.srh.noaa.gov/jetstream/tstorms/tornado.html

US Department of Commerce National Oceanic and Atmospheric Administration 2017, 'Tornado Definition', viewed 2 September 2017, https://www.weather.gov/phi/TornadoDefinition

US Department of Commerce National Oceanic and Atmospheric Administration 2017, 'Types of Thunderstorms – Supercell Thunderstorms', viewed 15 September 2017, http://www.srh.noaa.gov/jetstream/tstorms/tstrmtypes.html

US Department of Commerce National Oceanic and Atmospheric Administration 2017, 'What is a Supercell?' viewed 8 August 2017, https://www.weather.gov/ama/supercell

US National Park Service, Uploaded 6 October 2022, https://www.nps.gov/grte/learn/education/classrooms/upload/Weather-Lore-Sayings.pdf

Watts, A 2014, *The Weather Handbook*, 3rd edn, Bloomsbury Publishing, Sydney.

Permissions

Copyright permissions have been obtained for all photographs and figures reproduced in this book, as detailed below.

Figure 2 – Lightning bolt during storm. Tallahassee, Florida, U.S. Credit: National Oceanic and Atmospheric Administration/Department of Commerce. Source: NOAA Legacy Photo, Image ID wea00666, NOAA National Weather Service Collection

Figure 3 – Thunderstorm over Watson Lake, Prescott, Arizona, U.S. Credit: Photographer Bob Larson, and National Oceanic and Atmospheric Administration/Department of Commerce. Source: Image ID con00001, NOAA National Weather Service Collection.

Figure 5 – Intense cloud to ground lightning over southern Lake Michigan, Chesterton, Indiana, U.S. Photo #19: Credit: Photographer Nick Schrader, and National Oceanic and Atmospheric Administration/Department of Commerce. Source: Image ID: con00362, NOAA National Weather Service Collection

Figure 6 – Updraft tilts with wind shear. Photo #20: Credit: National Oceanic and Atmospheric Administration/Department of Commerce. Source: https://www.weather.gov/ilx/swop-springtopics

Figure 7 – Twin towering cumulus. Credit: National Oceanic and Atmospheric Administration/Department of Commerce, and Collection of Dr Bill Hooke, NOAA (ret.). Source: Image ID wea03530, NOAA National Weather Service Collection.

Figure 8 – Cumulonimbus with magnificent anvil. Photo #26: Credit: National Oceanic and Atmospheric Administration/Department of Commerce. Source: Image ID wea02023, NOAA National Weather Service Collection

Figure 9 – Lightning before rain seen from back yard in Rochester, New York, in the town of Greece. Credit: Photographer Patty Singer, and National Oceanic and Atmospheric Administration/Department of Commerce, NOAA. Source: Image ID con00034, NOAA National Weather Service Collection

Figure 10 – Gale-force winds lash huge ocean waves into a violent stretch of water. Drake Passage between Atlantic Ocean and Pacific Ocean. Credit: Photographer Jason Edwards. Source: http://www.alamy.com/stock-photo-gale-force-winds-lash-huge-ocean-waves-into-a-violent-stretch-of-water-35914522.html

Figure 12 – Cold front with squall line. Northern Adriatic Sea, Italy. Credit: Photographer Marco Korosec, and National Oceanic and Atmospheric Administration/Department of Commerce. Source: http://www.photolib.noaa.gov/nws/

Figure 13 – Sailboat approaching squall line. Pamlico Sound, North Carolina, U.S. Credit: Photographer Michael Halminski, Waves, North Carolina, and National Oceanic and Atmospheric Administration/Department of Commerce. Source: Image ID wea02202, NOAA National Weather Service Collection

Figure 14 – Overshooting top cumulonimbus cloud. Credit: National Oceanic and Atmospheric Administration/Department of Commerce. Source: http://www.photolib.noaa.gov/htmls/wea00106.htm

Figure 15 – Rain shaft to left, rain-free base of severe thunderstorm to right. Key West, Florida, U.S. Credit: Photographer Lieutenant Debora Barr NOAA Corps, and National Oceanic and Atmospheric Administration/Department of Commerce. Source: Image ID wea02191, NOAA National Weather Service Collection

Figure 16 – Severe Thunderstorm at Era Beach south of Sydney, Australia. The wall cloud (pedestal cloud) is lowered beneath the base of a cumulonimbus cloud. Picture also shows rain-free base that increases storm's longevity. Credit: Photographer Bruce Cooper, and Australian Weather Calendar. Source: Pentago, C, n.d., 'Australia Climate: Coping with severe thunderstorms', *Splash magazines*, http://www.lasplash.com/publish/Home_134/australia-thunderstorms.php

Figure 17 – Stylised tornadic supercell thunderstorm. Reproduced by permission of Bureau of Meteorology, © 2017 Commonwealth of Australia. Source: http://www.bom.gov.au/storm_spotters/handbook/images/fig1.jpg

Figure 19 – Aerial view of a supercell thunderstorm. Photograph taken looking northeast over eastern Kansas, U.S. Credit: Photographer T. Theodore Fujita, and National Oceanic and Atmospheric Administration/Department of Commerce. Source: http://www.spc.noaa.gov/misc/AbtDerechos/supercells.htm

Figure 20 – Intense updrafts produce a rain-free cloud base in a supercell. Credit: UCAR and NSSL. Source: Image ID nssl0409 http://www.photolib.noaa.gov/htmls/con00306.htm, NOAA National Severe Storms Laboratory (NSSL) Collection

Figure 21 – Waterspout in the Gulf of Mexico photographed from the NOAA ship *Rude*. South of Cameron, Louisiana, Gulf of Mexico. Credit: Collection of Commander Grady Tuell, NOAA Corps, and National Oceanic and Atmospheric Administration/Department of Commerce. Source: http://www.photolib.noaa.gov/htmls/wea00310.htm

Figure 22 – Twin violent (EF4) tornadoes, Wisner, Nebraska, U.S. Credit: Photographer Ethan Schisler, and National Oceanic and Atmospheric Administration/Department of Commerce. Source: Image ID con00002, NOAA National Weather Service Collection.

Figure 28 – Lightning shoots up the updraft and anvil of tornadic supercell at night with car light trails. Contributor: Cultura Creative (RF) / Alamy Stock Photo. Image ID: GDP9TE.

Figure 31 – Hurricane viewed from satellite. Credit and source: https://pixabay.com/en/hurricane-earth-satellite-tracking-92968/ (public domain)

Figure 32 – How firestorms form. Credit: National Oceanic and Atmospheric Administration/Department of Commerce and NASA. Source: https://scijinks.gov/firestorm/

Figure 37 – Swell lines in the Pacific. Credit: Alamy Photo. Source: National Geographic Creative.

Figure 38 – Swell from a distant storm apparent on the sea surface. Credit: Photographer John Bortniak, and National Oceanic and Atmospheric Administration/Department of Commerce. Source: Image ID corp2756, NOAA National Weather Service Collection.

Figure 39 – Swell. Source: https://pixabay.com/en/sunrise-on-the-sea-romantic-275274/ (public domain)

Figure 40 – Virga. View SW from Flat Top Mountain, North Carolina, U.S. Credit: Photographer Grant W. Goodge, and National Oceanic and Atmospheric Administration/Department of Commerce. Source: Image ID wea00057, NOAA National Weather Service Collection

Figure 41 – Cirrus at sea. Credit: Photographer Lieutenant Elizabeth Crapo, NOAA Co, and National Oceanic and Atmospheric Administration/Department of Commerce. Source: Image ID wea03609, NOAA National Weather Service Collection

Figure 42 – Cirrocumulus cloud, Michigan, Grand Rapids, U.S. Credit: Photographer Ralph F. Kresge, and National Oceanic and Atmospheric Administration/Department of Commerce. Source: #1060 Image ID wea00058, NOAA National Weather Service Collection, http://www.photolib.noaa.gov/htmls/wea00058.htm

Figure 43 – Sun halo and cirrostratus clouds at sunset. Contributor: Craig Joiner Photography / Alamy Stock Photo. Image ID: FT-KX8X.

Figure 44 – Mackerel sky of altocumulus clouds over the eliptic crater of Erta Ale volcano. Contributor: RWEISSWALD / Alamy Stock Photo. Image ID: FP8E6X.

Figure 45 – Data showing no recent slowdown in global warming. Credit: National Oceanic and Atmospheric Administration/ Department of Commerce. Source: http://www.noaanews.noaa. gov/stories2015/images/no%20slow%20down%20in%20global%20 warming.jpg

Figure 46 – Dust storm approaching Stratford, Texas, U.S. Credit: National Oceanic and Atmospheric Administration/Department of Commerce and George E. Marsh Album. Source: Image ID theb1366, NOAA National Weather Service Collection

Figure 47 – The 2016 snowstorm in Washington D.C. ranked as a category 4 storm on the NESIS scale. Credit: Photographer Joe Flood, and National Oceanic and Atmospheric Administration/ Department of Commerce. Source: http://www.noaanews.noaa. gov/stories2016/012816-noaa-ranks-january-2016-blizzard-category-4-on-the-northeast-snowfall-impact-scale.html

Figure 49 – Storm cloud near Bundeena, Australia. Credit: Photographer Bethany Needham.

Figure 50 – Lifeguard chair on beach. Contributor: Tetra Images / Alamy Stock Photo. Image ID: BGYHMF.

About the Author

Robert Ellis has a BSc (Hons) from Sydney University and has completed postgraduate qualifications at three Australian universities. He has published internationally in the US. He served for seven years as a Flight Lieutenant with a Permanent Commission in the RAAF. He taught aircrew at the Central Flying School and the School of Air Navigation. He taught on the first weapons systems course in Australia. He was a project manager in Technical and Further Education for 25 years.

Robert is a scientist and storm expert who has been referred to in recent years as a storm chaser in a weekend magazine of News Ltd.

Third edition

As many as 500,000 people worldwide may die in large storms each year. Traditional weather forecasts can currently only give around 13 minutes' lead time for tornadoes spawned by supercell thunderstorms. The Tornado Early Warning Rule presented for the first time in this book is ground-breaking, as you can now have at least 5 hours' early warning about a tornadic supercell thunderstorm. The extra warning time will save many lives.

Robert Ellis' book shows ordinary people how to predict a storm long before it is even visible to radar or satellite. Many lives can be saved by using the simple rules explained in the book.

Predicting Storms covers practical information such as whether you can walk to work, or if there will be a storm or rain in your area within the next hour or two.

All types of storms are covered in the book: Severe Thunderstorms, tornadic supercell thunderstorms, cyclones, hurricanes, typhoons, extratropical cyclones, tropical storms, tornadoes, firestorms, weather bombs, windstorms, dust storms, and snowstorms.

Whether you are a general reader, a surfer, a weather watcher, a storm-spotter, or a storm-chaser, *Predicting Storms* will give you the tools to predict all storms confidently.

ISBN 978-0-6481072-6-2

9 780648 107262 >

www.ingramcontent.com/pod-product-compliance
Lightning Source LLC
Chambersburg PA
CBHW040859210326
41597CB00029B/4900